Lost History

EXPLORATIONS IN NUCLEAR RESEARCH, VOL. 3

A precursor to modern research in low-energy nuclear reactions (1912-1927)

Steven B. Krivit

Edited by Michael J. Ravnitzky

Pacific Oaks Press
SAN RAFAEL, CALIFORNIA

Lost History: Explorations in Nuclear Research, Vol. 3
Copyright © 2016 by Steven B. Krivit

All rights reserved. No part of this book may be reproduced in whole or in part without written permission from the publisher, except by reviewers who may quote brief excerpts in connection with a review in a newspaper, magazine, or electronic publication; nor may any part of this book be reproduced, stored in a retrieval system, or transmitted in any form or by any means without permission from the publisher.

Pacific Oaks Press / *New Energy Times* / www.newenergytimes.com
369-B 3rd St. #556, San Rafael, CA 94901
Library of Congress Control Number: 2015916494

Krivit, Steven B., author.
　Lost history / Steven B. Krivit ; editors, Michael
Ravnitzky, Cynthia Goldstein, Mat Nieuwenhoven.
　　pages cm -- (Explorations in nuclear research ; vol. 3)
　　Includes bibliographical references and index.
　　LCCN 2015916494
　　ISBN 978-0-996886406 (hbk.)
　　ISBN 978-0-996886413 (pbk.)
　　ISBN 978-0-996886420 (Kindle)
　　ISBN 978-0-996886437 (ePUB)

　　1. Low-energy nuclear reactions--Research--History.
　2. Nuclear reactions--Research--History.
　I. Ravnitzky, Michael, editor. II. Goldstein, Cynthia, editor. III. Nieuwenhoven, Mat, editor. IV. Title.
　V. Series: Krivit, Steven B. Explorations in nuclear research ; v. 3.

QC794.8.L69K755 2016　　539.7'5
QBI16-600062

Cover design: Lucien G. Frisch (Photograph: © Jahoo | Dreamstime.com)
Interior design template: Book Design Templates Inc.
Typeset in Crimson 11 pt., designed by Sebastian Kosch
Editors: Michael Ravnitzky (Developmental Editor), Cynthia Goldstein (Copy Editor), Mat Nieuwenhoven (Technical Editor)
Index: Laura Shelley

Also by Steven B. Krivit

Hacking the Atom: Explorations in Nuclear Research, Vol. 1 (2016)

Fusion Fiasco: Explorations in Nuclear Research, Vol. 2 (2016)

Nuclear Energy Encyclopedia: Science, Technology, and Applications,
Steven B. Krivit, Editor-in-Chief; Jay H. Lehr, Series Editor,
Wiley Series on Energy (2011)

American Chemical Society Symposium Series: Low-Energy Nuclear Reactions and New Energy Technologies Sourcebook (Vol. 2),
Jan Marwan, Steven B. Krivit, editors (2009)

American Chemical Society Symposium Series: Low-Energy Nuclear Reactions Sourcebook (Vol. 1),
Jan Marwan, Steven B. Krivit, editors (2008)

The Rebirth of Cold Fusion: Real Science, Real Hope, Real Energy,
by Steven B. Krivit and Nadine Winocur (2004)

www.NewEnergyTimes.com (since 2000)

Illustration by André Castaigne, "M. Pierre and Mme. Marie Curie Finishing the Preparation of Some Radium." First published in McClure's Magazine, Vol. XXII, No. 1, Nov. 1903

Belief in the simplicity of nature is not logic but faith pure and simple. It is one of those insidious and dangerous tacit assumptions which often creep into scientific theories. Tacit assumptions are dangerous because they are usually made unconsciously, so that they appear to be self-evident truths and prevent our harboring the shadow of a doubt of their insidious character.

Joseph William Mellor (1873-1938) (Mellor, 1923, 1)

To Jess
Steven B. Krivit

To my sons Nathan and Max Ravnitzky
Michael J. Ravnitzky

Acknowledgments

Acknowledgments for the Series

Several key people have assisted me in the past few years on these books. They are my personal heroes. First is my copy editor, Cynthia Goldstein, who has been part of my team since 2004. Cynthia is my secret weapon who, year after year, helps make my writing intelligible.

Three other people composed the core team that made these books possible. Initially, they responded to my request to critique each chapter as Cynthia and I produced them. All of them exceeded my expectations in each of their unique contributions. None of them is a scientist, but all of them have some scientific background or technical training. I asked them to find all possible flaws and errors, and indeed they brought much to my attention. Of course, if any errors remain, they are my fault alone.

Michael Ravnitzky, in Maryland, accepted my invitation to critically review the draft chapters. After I convinced him that I really did welcome every critique he had to offer, he provided an invaluable outpouring that contributed immensely to the development of the books. Michael was a dream to work with. I am forever in his debt for his insights, wisdom, and his relentless, constructive and diplomatic suggestions, and so many other thoughtful contributions.

Mat Nieuwenhoven, in the Netherlands, has an eye for details like I have not imagined possible. His dedication to this project and helping make it as technically accurate as possible was heroic.

Lucien G. Frisch, in Germany, could see and understand the larger vision of these books from the very beginning, as soon as he read the first completed chapter. Sometimes, he could see the purpose of these books more clearly than I did, as I was on occasion too close to it to see it myself. I thank him not only for his visionary guidance but also for his brilliant artistry that graces the covers of these books.

I wish to extend my heartfelt appreciation and gratitude to the following reference librarians, who have provided me with so much support for all three volumes: my hero Randy Souther, at the Gleeson Library, Geschke Center, University of San Francisco; Lorna Whyte, Diane Delara and Pam Klein at the San Rafael Public Library; and Lorna

Lippes, an independent researcher. I owe much appreciation to Libby Dechman, at the Library of Congress, for her help in establishing the new cataloging subject heading "low-energy nuclear reactions." Thanks to all of you; I have a newfound and deep appreciation for reference librarians and public libraries.

I wish to thank Larry, Dee, Mirabai and Tracy for their assistance in helping me see the path on which I was travelling. Thank you, Flori, Jessica, Christine, Al and Sean, for your encouragement and support. Last but not least, many thanks to Sophie Wilson for crucial conversations that have helped me understand how to tell these stories.

Acknowledgments for This Volume

As Chapter 9 reveals, this book would have been impossible without the historical research of Robert A. Nelson and his book *Adept Alchemy*. Robert, thank you for embarking on your strange and wonderful journey and leaving behind the bread crumbs for me and others to follow.

I thank Carmen Giunta, professor of chemistry at Le Moyne College, in Syracuse, New York, who generously read and critiqued early drafts of Chapters 1-14 and Chapter 22.

I would like to extend my deepest appreciation to Toni Naebauer and Mat Nieuwenhoven for their assistance in translating the 1926 Paneth and Peters paper from German to English. Thank you to Maxime Petiau for translating the 1927 Paneth, Peters, and Günther paper. Thanks to Silke Merchel for her organizational help with the Paneth papers.

Table of Contents

Introduction xiii

1. Deceit, Falsehood, and Delusion 1
   Elemental Transmutation Research Is Dismissed As Fraud, Folly, and Failure

2. Alchemy, the Precursor to Modern Science 5
   Around the World, Mystics, Magicians and Madmen Defy Science Authorities

3. A Strange Glow; An Unusual Shadow 21
   Photography Opens the Door to the Discovery of Radiation (1896)

4. Atomic Boot Camp 29
   Basic Training in Nuclear Science for the Rest of Us

5. Don't Call It Transmutation! 35
   Soddy Envisioned It, Rutherford Avoided It: the First Observation of Natural Nuclear Disintegration (1902)

6. The Roots of Nuclear Energy 47
   Mysterious Radium in Marie and Pierre Curie's Lab Burns Brightly but Is Not Consumed (1903-1904)

7. Yes, It Is Transmutation! 51
   Soddy and Ramsay Destroy a 100-year-old Assumption and Prove Natural Elemental Transmutation (1903)

8. Giving Credit Where It's Not Due 57
   Soddy Describes the Disintegration Theory; Rutherford Gets the Credit and the Nobel Prize (1903-1904)

9. From LSD to Alchemy 71
   One Man Found It: the Lost Trail of Early Transmutation Research

10. Ramsay Breaks the Ice 77
   Chemist's Attempt at Man-Made Transmutation Offers Lessons on the Scientific Method (1907)

11. Soddy: Keen Observer and Visionary 95
   Chemist Foresees the Potential of Nuclear Power and the Value of Transmutation (1908)

12. Visionary Scientists — or Not? 105
   Ramsay Foresees Coal Decline; Rutherford Fails to See Atomic Energy

13. Three Mysterious Rare Gases 111
   Finally, Man-Made Nuclear Transmutations, but Physicists Are Not Convinced (1908-1914)

14. Out of Gas 139
   After Dramatic Headlines, Interest in Chemists' Transmutations Fizzles (1913-1927)

15. Rutherford Shatters the Atom 147
   The First Man-Made Disintegration of a Stable Atom Generates an Abundance of Alchemical Myths (1919)

16. Golden Rumors 157
   Rumors Spread That Rutherford Turned Lead into Gold and Germans Manufactured Artificial Gold (1920-1922)

17. Hotter Than the Sun 167
   Danger: Exploding Wires!

18. Exploding Wires in Chicago 175
   Wendt and Irion Enter Forbidden Land of Transmutation, Explode Tungsten and Make Helium (1922)

19. Flawless Yet Discredited Experiment 199
   Other Scientists Suggest That Their Own Failures Disprove Wendt and Irion's Successes (1922-1924)

20. Accidental Alchemist 213

Miethe Finds Trace Amounts of Gold in the Residue of Mercury Vapor Photography Lamps (1924-1926)

21. Golden Journey 237
Nagaoka Finds Visible Specks of Gold While Investigating Mercury Isotopes (1924-1925)

22. The Death of Modern Alchemy 251
A High-Profile and Highly Publicized Failed Replication Puts an End to the Early Transmutation Era (1925-1927)

23. Recovering a Milestone in Scientific History 273
Credit for the First Confirmed Transmutation Belongs Not to Rutherford but to His Student (1925)

24. A Bold Advance, A Hasty Retreat 297
Paneth and Peters Claim Helium by Transmutation, Then Lose Their Courage and Try to Retract (1926-1927)

25. Asleep for 60 Years 315
Summary of the New or Forgotten Contributions to the Scientific Literature Shown in This Book

Appendix A - Emissions From Spontaneous Radioactivity 321
Appendix B - Helium Permeation in Metals Analysis 323
Appendix C - Cathode-Ray Tube Research/Rare Gases 325
Appendix D - Rutherford's Nitrogen Disintegration 327
Appendix E - Timeline of Key Events in Early Nuclear History 329
Appendix F - Definition of Low-Energy Nuclear Reactions 337
Glossary of Scientific Terms 339
Bibliography 347
Index 359
About the Author 381

Introduction

In the early 20th century, when fundamental discoveries and breakthroughs were made in physics and chemistry, a parallel track of scientific exploration and discovery took place between 1912 and 1927 that has been lost and forgotten — until now.

Scientists performing experiments in defiance of prevailing theory at the time used relatively simple benchtop apparatus that transmuted elements without using a radioactive source. In some cases, they produced noble gases and, in other cases, precious metals.

At the time, critics gave a variety of reasons to dismiss the claimed results as erroneous. However, for most of the experiments, no critics identified a specific error of protocol, a mistake in the data analysis, or an unstated assumption by the researchers who reported these transmutations.

Historians wrote off these experimental results primarily because some scientists failed in their replication attempts. Secondarily, historians have assumed that critics' guesses about the presumed faults were correct. (Mellor, 1923)

Neither the interpretation of the failed replication attempts nor the guesses about the errors were correct. This is the first book that provides critical analyses of the original published scientific papers of the transmutation experiments performed between 1912 and 1927.

The chapters in this book reveal the story of these experiments and provide significant insights about our understanding of the history of physics, chemistry and nuclear science.

History as We Know It

The lost history period took place between two well-known eras of nuclear research. Preceding the lost history, in the first decade of the 20th century, scientists learned a great deal about atomic reactions induced by alpha particles emitted from passively decaying radioactive elements. After the lost history period, in the 1930s, scientists began using devices to accelerate ions and induce nuclear disintegrations with higher energies and greater control.

INTRODUCTION

The research during both well-known eras was generally consistent with scientists' theoretical knowledge at the time. They understood that the emissions from alpha particles, from naturally decaying radioactive elements or from particle accelerators carried sufficient energy to induce nuclear reactions.

Alchemy?

On the other hand, the transmutation experiments reported between 1912 and 1927 used no radioactive sources or particle accelerators. Instead, the experiments used a variety of relatively low-energy stimuli.

The results made no sense according to theory at the time, and the research looked suspiciously like alchemy, an ancient craft that had no place in respectable science.

At the time, this research was widely known both in scientific circles and among the general public, despite its association with alchemy. Although many scientists who heard about the work doubted that it was real, journals nevertheless published scientific papers on the research, and newspapers eagerly informed the public.

The work was reported in popular media, such as the *New York Times* and *Scientific American*. Scientific papers were published in the top professional journals of the day, including *Physical Review* and *Nature*. Prominent scientists in the U.S., Europe, and Japan and even Nobel Prize winners were active in the research.

By the 1930s, the entire body of research and results was dismissed as false and was left out of history books for nearly a century.

This book scrutinizes the most significant claims and counterclaims made during this lost history. Research for this book included the analysis of more than 140 original scientific papers and more than 200 news articles from the first three decades of the 20th century.

Three Books

This is the third book in a three-book series. Each book stands alone, and covers a distinct period of scientific exploration. They are being published in reverse-chronological order.

INTRODUCTION

- *Hacking the Atom: Explorations in Nuclear Research, Vol. 1 (1990-2015)*
- *Fusion Fiasco: Explorations in Nuclear Research, Vol. 2 (1989-1990)*
- *Lost History: Explorations in Nuclear Research, Vol. 3 (1912-1927)*

The 1912-1927 era of transmutation research is best understood with the insights from the experimental research discussed in Vol. 1, *Hacking the Atom*.

The sources used in this book are primarily published scientific papers. For this reason, this book is geared toward a more technical and academic audience than the other two books in this series. Here are some highlights from this book:

Anomalous Production of Noble Gases

From 1912 to 1914, several independent researchers detected the production of the gases helium-4, neon, argon, and an as-yet-unidentified element of mass-3, which we now identify as tritium. Two of these researchers were Nobel laureates. In 1922, two chemists at the University of Chicago also reported the production of helium-4.

Wendt and Irion's Synthesis of Helium

In 1922, two chemists at the University of Chicago, Gerald L. Wendt and Clarence E. Irion, synthesized helium using the exploding electrical conductor method. Despite doubts and criticism, no one unambiguously identified any error in their 21 successful experiments.

Nuclear evidence from exploding electrical conductor experiments was confirmed 80 years later by researchers at the Kurchatov Institute in Russia. (Urutskoev, 2002)

Anomalous Production of Gold, Platinum, and Thallium

In 1924, a German scientist accidently found trace amounts of gold and possibly platinum in the residue of mercury vapor lamps that he had been using for photography. A year later, scientists in Amsterdam repeated a similar experiment, but starting with lead, and observed the production of mercury and the rare element thallium. The same year, a prominent Japanese scientist, in a different kind of experiment, reported observing the production of gold and something that had the appearance of platinum. Newspapers reported that he toured the world showing people the gold he had made in the laboratory. No reports of challenges to his claim appear to exist, at least in English-language references.

Paneth and Peters' Hydrogen-to-Helium Transmutation

In 1926, German chemists Friedrich Adolf (Fritz) Paneth and Kurt Gustav Karl Peters pumped hydrogen gas into a chamber with finely divided palladium powder and reported the transmutation of hydrogen into helium. Paneth was at first very proud of his and Peters' achievement. Paneth claimed that they were the first scientists to perform a nuclear transmutation. He dismissed the earlier 1912-1914 transmutation reports without thoroughly examining them and without clearly identifying any errors.

A year later, Paneth did an about-face: He worked hard to find explanations to dismiss his and Peters' helium-production claims. He was unable to completely explain away their results.

Correction to a Milestone in Scientific History

While doing research on this early transmutation era, I came across facts that contradict the depiction of an important milestone in scientific history. World-famous physicist Ernest Rutherford has been credited incorrectly with the first nuclear transmutation. Some historians, even Rutherford scholars, call it his greatest achievement.

Not only was he preceded by other researchers in the 1912-1914 era, but also the experiment that has been attributed to him, transmuting nitrogen to oxygen, was in fact performed by Patrick Maynard Stewart Blackett, a research fellow who was working under Rutherford.

All but a few historians — and all known current Internet references as of 2015 — incorrectly credit this discovery to Rutherford.

Preparation for a Paradigm Shift

The impetus for my inquiry into this lost history came from my curiosity about the 1989 "cold fusion" claim by two electrochemists, Martin Fleischmann and Stanley Pons, at the University of Utah. The chemists observed valid anomalous experimental phenomena but made

INTRODUCTION

an incorrect interpretation of the data, claiming it was produced by a fusion reaction.

There is no experimental evidence to support the claim of "cold fusion"; nor is there a viable theory to explain it as fusion, were supportive experimental data to exist.

There is, however, an abundance of experimental evidence as well as a non-fusion theory that offers a viable explanation for the phenomena. Electroweak interactions appear to be the missing link to understanding these mysterious reactions.

Hacking the Atom: Explorations in Nuclear Research, Vol. 1 discusses the research that took place in the field from 1990 to 2015, as well as the non-fusion theory.

Fusion Fiasco: Explorations in Nuclear Research, Vol. 2 discusses the 1989-1990 "cold fusion" history.

With the benefit of hindsight, and in light of modern research and theory, this lost history takes on new meaning and a new place in the history of science. It is also a remarkable precursor to the low-energy nuclear research (LENR) that began in the 1980s.

In this latter era, researchers again reported a wide variety of nuclear transmutations at low energies, including noble gases and precious elements. The data were taken with modern electron microscopy in recognized laboratories around the world.

This 100-year continuum reveals the emergence of a new field of science that belongs to neither physics nor chemistry but is a hybrid of the two.

Welcome to the Journey

I have independently investigated and reported on this subject for 16 years. I invite scientists and non-scientists alike to join me on this journey of scientific exploration and discovery. It is my pleasure to share this adventure with you now.

Steven B. Krivit
San Rafael, California
Sept. 1, 2016

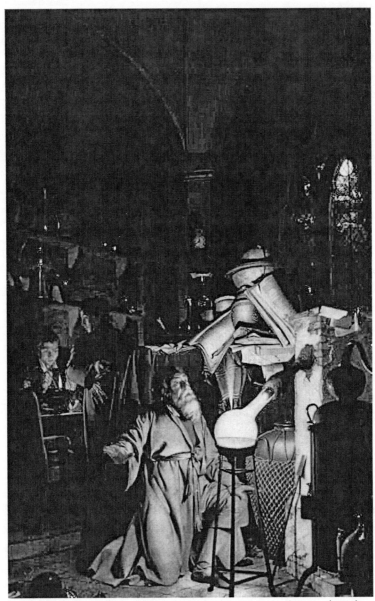
Portion of painting of alchemist by Joseph Wright of Derby (1771)

CHAPTER 1

Deceit, Falsehood, and Delusion

Elemental Transmutation Research Is Dismissed As Fraud, Folly, and Failure

Our journey begins in the early 20th century. Our scribe is Joseph William Mellor, a ceramic chemist who edited encyclopedic compendiums of scientific research in organic and inorganic chemistry.

His foremost legacy is preserved in a 16-volume set of textbooks. The series is called *A Comprehensive Treatise on Inorganic and Theoretical Chemistry*. Most volumes are a thousand pages. There is no question as to the breadth and depth of Mellor's knowledge of chemistry.

In his Volume 4, originally printed in 1923 and reprinted in later years, Mellor discussed the transmutation claims made between 1912 and 1927. Like nearly all scientists of his day, Mellor viewed the transmutation claims as alchemy, which he disdained:

> J.J. Pontanus (1520) complained that after traveling through many countries to examine the claims of the adepts, he found many deceivers, but no true philosophers; and N. Lemery, that "they professed an art the beginning of which was deceit, the progress of which was falsehood, and the end beggary."
>
> Alchemy thus fell into disrepute, for it seemed as if its claims could be established only by chicanery and fraud. ... At one period, however, the majority of alchemists did seek

to make gold cheaply with the sole object of gaining untold wealth. Failure or delusion was inevitable. Accordingly, the alchemist often misrepresented the truth and that degenerated into a charlatan and imposter, pretending, with vulgar frauds, that he had succeeded. (Mellor, 1923, 148)

Mellor summarized most of the significant transmutation claims between 1912 and 1923:

The alchemist's dream of transmutation is very little nearer realization today than it was a thousand years ago, for no one has yet really succeeded in transmuting one chemical element into another other than by speculative argument. There is no unimpeachable evidence of a single transmutation of one element into another. ...

J. J. Thomson has stated that all his efforts to decompose atoms by cathode rays or positive rays have failed to produce any conclusive evidence of a transformation. True enough, a few radioactive elements — radium, actinium, polonium, uranium, and thorium — seem to have been discovered in nature, and they are usually stated to be changing spontaneously from one elemental form to another; but no process known to man is able to accelerate or retard, stop or start the metamorphosis. No element has yet been broken down into a simpler substance by a process controllable by man. In the words of Francis Bacon, *natura enim non nisi parendo I vincitur* — nature to be conquered must be obeyed.

The alleged transmutation of copper into lithium and sodium by A.T. Cameron and W. Ramsay has been denied by M.S. Curie and E. Gleditsch and by E.P. Perman — the lithium and sodium were derived from the vessels used in the work; the production of neon from radium emanation by W. Ramsay, and W. Ramsay and A.T. Cameron has been denied by E. Rutherford and T. Royds — the neon was derived from the air which had not been excluded from the apparatus; and the formation of carbon dioxide by the action of radium emanations on solution of thorium and zirconium by W. Ramsay and F.L. Usher has been called in question by E. Rutherford — the carbon appears to have been derived from the grease used in lubricating the stopcocks.

The alleged transmutation of hydrogen into neon, by W. Ramsay, J.N. Collie and H.S. Patterson, and I. Masson, by the action of a stream of cathode rays on hydrogen is considered by J.J. Thomson to be a mal-inference, since the neon is thought to be derived from that originally occluded by the electrodes, or glass vessel, and which is expelled by the bombardment of the cathode rays, but which cannot be removed by the mere application of heat. R.J. Strutt, T.R. Merton, A.C.G. Egerton, and A. Piutti and E. Cardoso could not verify the alleged conversion of hydrogen into neon. (Mellor, 1923, 149-150)

Mellor and other scientists in 1923 knew that changes to atomic nuclei required high energies:

It has been pointed out that the formation of, say, gold from a metal atomically lighter, say tin, would require the expenditure of so much energy that even if the transformation were accomplished, it could not be a successful commercial process for the production of gold. On the other hand, the formation of gold from an atomically heavier metal, say lead, would liberate such an enormous amount of energy that the gold would be but an insignificant by-product, for the energy liberated during the process would have an enormously greater value than the metal. (Mellor, 1923, 150)

Mellor and his contemporaries knew that alpha particles randomly emitted from passively decaying radioactive elements could provide the required energy to cause atomic disintegrations. But not until 1925 did physicists observe the first man-made, alpha-induced elemental transmutation. In 1932, physicists used devices to accelerate ions and induce nuclear disintegrations, and thereafter, transmutations by high-energy physics were easily accepted as *bona fide* scientific phenomena.

Mellor's perspective was limited by the prevailing theoretical understanding of atomic science. It would take 82 years for two theorists, Allan Widom (Northeastern University) and Lewis Larsen (Lattice Energy LLC), to propose a theory based on electroweak interactions that offered a feasible explanation of the then-inexplicable phenomena. (See the book *Hacking the Atom: Explorations in Nuclear Research, Vol. 1* for more information.)

Disbelief of the reported results was understandable in 1923, and denial of the experimental results was far more logical than acceptance. It made more sense at the time to deny the data, however empirically obtained, by making guesses or assuming there had been unidentified errors, as Mellor and other scientists did.

It made more sense to them to dismiss the claims by mentioning that other scientists "could not verify" the claims when attempting a replication.

Mellor's quote, displayed in the epigraph and shown here, becomes ever more useful in understanding why these good experiments were dismissed by even the most knowledgeable scientists:

> Belief in the simplicity of nature is not logic but faith pure and simple. It is one of those insidious and dangerous tacit assumptions which often creep into scientific theories. Tacit assumptions are dangerous because they are usually made unconsciously, so that they appear to be self-evident truths, and prevent our harboring the shadow of a doubt of their insidious character. (Mellor, 1923, 1)

These transmutation experiments represented — and still represent — nothing less than a challenge to our assumptions about nature. These experiments show us that there is something new to learn in science. They remind us that nuclear physics is more than strong-force fission and fusion. They teach us that another of the four fundamental forces, weak interactions, heretofore with limited practical application, may portend new worlds of science and technology.

CHAPTER 2

Alchemy, the Precursor to Modern Science

Around the World, Mystics, Magicians and Madmen Defy Science Authorities

The well-known early period of atomic and nuclear science took place between 1895 and 1930. During the later part of this period, scientists were pursuing a parallel track of research that has, almost without exception, been omitted from today's accounts of science history. In order to provide a basic foundation for all readers, the first chapters (2-8) review the scientific developments during the earlier part of this period.

The birth of atomic (later known as nuclear) science took place around the turn of the 20th century, between 1895 and 1930. During the first three decades of the 20th century, there were no hard boundaries between chemists and physicists in the domain of atomic and nuclear research. At the time, they simply called the research "radioactivity." The research was shared equally by both disciplines, as we will see in the coming chapters.

Physicists eventually explored and identified the nucleus and atomic science gave way to nuclear science. Physicists took the research under their wing, and in the mid-1930s, the research became known as "nuclear physics." Starting in the 1930s, physicists navigated the new subatomic world more effectively than chemists by using the methods and tools of particle physics. When this shift occurred, chemists were no longer viewed, even by themselves, as equal players in nuclear research.

In the 1940s, when nuclear fission-based military and energy programs began, the term "nuclear chemistry" came into use to designate the chemistry-related aspects of that work. Nuclear chemistry was essential to aspects of the atomic bomb program and subsequent research on transuranic elements. Around this time, nuclear chemists typically worked at laboratories where nuclear physicists were working with particle accelerators or nuclear reactors.

These devices did not exist in the first three decades of the 20th century; chemists, as well as physicists, performing atomic research at that time worked with much simpler tools and techniques.

Until the late 1920s, before the field of radioactivity (or atomic science) became known as nuclear physics, physicists and chemists routinely worked much more closely than they do today. In a few short years during and after the turn of the century, chemists and physicists worked collaboratively and made a series of monumental discoveries in rapid succession, each one building on the previous.

Yet the new world of atomic science did not arrive without the fear of the return of alchemy coming along with it, a subject which scientists reviled. Long before chemistry and nuclear physics were recognized, alchemy was often regarded as the domain of mystics, magicians and perhaps a few madmen.

According to Adam McLean, a historian of alchemy, there are three possible sources for the origin of the term. The first is from ancient Egyptian, transmitted through the Arabic word "al-khem," based on the Egyptian hieroglyphics. The other derivations are from Greek words.

Alchemy Throughout History

Alchemy goes back several thousand years and has its roots in Egypt, India and China. Robert Place gives a good description of alchemy in his book *Magic and Alchemy: Mysteries, Legends, and Unexplained Phenomena*:

> Alchemy is as confusing a subject as magic. If asked to give a meaning for the word, ancient alchemists would have

given as many definitions as there were alchemists. One thing that they could all agree on, however, was that the central purpose of alchemy is transmutation. Transmutation is an event in which one substance is changed into another. Alchemists believed that transmutation was possible, and they made it the main focus of their work. The most famous example is their belief that they could change lead, an inexpensive metal, into gold, one of the most valuable [metals].

Alchemists hoped to accomplish transmutation through the interaction of a magical catalyst called the Philosopher's Stone. The creation of [what they called] the Philosopher's Stone, therefore, became the central purpose for the great work of all alchemists. The Philosopher's Stone was said to be a mystical substance: a stone that is not a stone. It could cure any illness, prolong life indefinitely, and transform any metal into its highest state. Some alchemical texts focus on creating the stone through lab work, and these were the precursors of modern medicine and chemistry. Others, however, focus on alchemy as an internal mystical process that takes place in the mind of the alchemist. (Place, 2009, 71-73)

Alchemists knew how to separate metals from ore and make alloys of metals. They used a variety of other techniques including brewing, dyeing, gilding, perfume-making, making chemicals and reciting magic rituals for the dead. McLean lists on his Alchemy Web Site (alchemywebsite.com) 1,179 authors of 2,810 books written on alchemy before the year 1800. McLean says that, by the 15th century, 300 texts from China were closely related to the various alchemical traditions.

According to Place, Western alchemy had its roots in Egypt and dealt primarily with metals. Some ancient Greeks also developed the alchemical theories based on certain fundamental aspects of nature: attraction and repulsion, love and hate, and the four elements, earth, fire, air, and water. The Greek philosophies were influenced by Egyptian mystical religion and magic, and scholars have found ancient alchemical manuscripts written on papyrus from Alexandria, Egypt, several hundred years B.C.E.

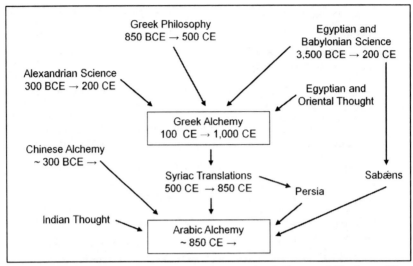

Table of alchemical tradition. Based on drawing by F. Sherwood Taylor in The Alchemist, Founders of Modern Chemistry *(1949)*

Alchemy in Europe

Alchemy entered Europe in the 12th century, according to Place. The Knights Templar were thought to be among the first Westerners to be acquainted with alchemy. During the Crusades, the Knights Templar had adopted teachings from a mystical pagan sect in the Islamic world. A few centuries later, when the Islamic empire in Spain lost territory to Christian rulers, scholars translated Arabic texts into Latin, and this propelled knowledge of alchemy to the rest of Europe. Soon, the alchemical influence permeated Europe, at first through Sicily, Spain and southern France.

When alchemy reached central Europe, it found its home in the castles of Europe's kings and queens. Records indicate that alchemists served their masters and provided them with newly made gold to line their masters' coffers, and when the alchemists failed, death was a common punishment.

CHAPTER 2 • 9

Fritz Paneth, a German chemist who performed transmutation experiments in the 1920s, gave a fascinating lecture at Cornell University on October 4, 1926, titled "Ancient and Modern Alchemy," that illuminates this history. The journal *Science* published his lecture later that month. Paneth found records suggesting that actual transactions based on manufactured gold took place, but on balance, there was very little evidence to support the legitimacy of the get-rich-quick scheme. Paneth believed the royal interest was based on wishful thinking, greed and the need for hard currency for wars and other wealth-consuming projects. Paneth's study provides a fascinating insight into history. Here is an excerpt:

> Official state papers of the sixteenth and seventeenth centuries make it clear that one of the important problems confronting a monarch or elector in Central Europe was to procure for his country an able alchemist who was expected to improve the financial status of the realm by transmuting base metals into valuable gold. It naturally followed that the alchemist was highly favored at court — so long as belief in his ability lasted. He was honored by the friendship of his sovereign and sometimes by elevation to nobility, and more than one of the crowned protectors of alchemy assisted personally in the experiments, so that he might convince himself of the correctness of the achievements of his alchemistic employee.
>
> The Emperor Rudolph II is reported to have [personally] worked with his alchemists. A visitor to the Hradshin, the beautiful castle of Prague, the residence of the emperor, may even today see the five or six little houses, with disproportionately large fireplaces, which were built, by Rudolph's command, close to his own palace and which were used by his "goldcooks." Rudolph appointed to a high position in his court Tycho Brahe, who, although usually referred to in the history of science as an astronomer, was perhaps chosen by Rudolph because he was also of high repute as an alchemist. This is evidenced by the fact that the

emperor provided him not only with an observatory but also with a laboratory for his chemical experiments.

In a more practical way, Henry VI of England supported alchemistical experiments. To aid in the payment of the debts of the state, he recommended to all noblemen, scholars and theologians the study of alchemy, and he conferred upon a company the privilege of making gold from base metals. This firm produced a metal (probably an alloy of copper and mercury) which had the appearance of gold, and from this, coins were stamped. History does not record whether King Henry believed that transmutation had actually been accomplished, but the careful Scotch were evidently skeptical, for the Scotch Parliament issued an order that this English "gold" should not be allowed to enter any of their ports or to cross their frontier.

The example given by the mightiest rulers of the time was imitated on a more modest scale by several of the smaller princes of Europe. Historical records tell us of one [person] who tried to obtain a first-class alchemist from his neighbor, first by kindness and then by force; of another prince who loaned his alchemist to another court for a definite period; and of treaties between two states in which alchemists were regarded as mere chattels. Many of the rulers of that time were such firm believers in alchemistical doctrines that a lawyer of the period advocated making disbelief in these theories a *crimen laesae majestatis.* But although the lords of the realms generously supported the experiments of their alchemists, the financial returns never seemed to equal the disbursements. [Readers] repeatedly find in the records that, at the end of the research, the sovereign lost his temper and that the alchemist, when hard-pressed to show his product of manufactured gold, was usually well-satisfied if he succeeded in escaping from the clutches of his former benefactor. If he failed to do so, he was severely punished and generally put to death. Showing the cruel humor of the times, it was a frequent joke to gild with [gold] tinsel the [hanging] gibbet on which the alchemist was to meet his end.

We read of a great number of such executions and of innumerable failures of experiments. The successful transmutation of some cheap material into gold was very seldom reported, and in every case, the transmutation, for some reason or other, could not be repeated: Either the alchemist had disappeared, or the stock of the "Philosophers' Stone," the miraculous powder which alone enabled him to accomplish "the great work," had been exhausted. The value of the gold that he claimed to have produced always amounted to a very small fraction of the money that had been spent upon him and his experiments. (Paneth, 1926, 410)

In his Cornell lecture in 1926, Paneth told the audience that the ancient efforts at transmutation were futile. (*Cornell Daily Sun*, 1926) He gave the impression that all attempts at ancient alchemy had failed. He also implied that attempts to perform elemental transmutations in the previous two decades had all failed. Meanwhile, around this time, Paneth and his colleague Kurt Peters, published papers claiming to have accomplished the very first successful elemental transmutation.

Greed

"By the 17th century," Place wrote, "interest in alchemy had peaked, and an unprecedented quantity of enigmatically illustrated alchemical books was published." One of the more fascinating among these was a book with no text, just pictures. The books covered a wide variety of topics including the use of non-herbal medicines and spiritual transformation.

Paracelsus, born Philippus Aureolus Theophrastus Bombastus von Hohenheim (1493-1541), Place wrote, was regarded as one of the greatest alchemists as well as the founder of modern medicine. A Swiss-German, he was also a physician, botanist, astrologer, and occultist. The spiritual quest had always been a part of alchemy from ancient times, but after Paracelsus, spirituality became more important to alchemists, as Place explained.

> These alchemists wanted to separate themselves from those [people] who were interested in alchemy only as a means to wealth. Solely materialistic alchemists were called puffers because of their impatient use of the bellows to keep the fire hot to speed up the process.
>
> Besides puffers, con men sought riches and fame through fraudulent claims of the successful transformation of base metal into gold, accomplished by trickery. These charlatans caused alchemy to fall into disrepute. In the 18th century, fraud [along with] the discrediting of alchemy's underlying theories by scientific discoveries caused alchemy to be reduced to a pseudoscience. (Place, 2009, 82)

The word puffery is commonly used to this day to denote marketing exaggeration. Now we know its origin!

Painting by Adrian van Ostade (1610-85), "The Alchemists," 1757 depicting an alchemist puffing his fire with a bellows.

Alchemy in India

The most unusual accounts of alchemy I've found reportedly took place in 20th century India, in Hindu temples. One of the most respected authorities on Indian alchemy was Vaidya Bhagwan Dash (1934-2015), author of *Alchemy and Metallic Medicines in Ayurveda*, published in 1986. Dash was an ayurvedic physician and scholar and author of more than 80 books on Ayurveda and Tibetan medicine as well as a Sanskrit scholar. Besides having other responsibilities, he was the deputy adviser in Ayurveda to the government of India in the Ministry of Health and Family Welfare and was a consultant in traditional medicine for the World Health Organization.

Among other lesser-known events, two prominent alchemical events took place in two prominent temples in India. They were recorded in Sanskrit inscriptions chiseled in marble plaques on pillars in the temples, built in the early 20th century by the Birla family, renowned Indian industrialists.

Lakshmi Narayan Temple (popularly known as Birla Temple) in New Delhi. Photo credit: Dash

One of the temples is in Banaras, also known as Varanasi, one of the oldest continuously inhabited cities in the world. The other is in New Delhi.

Yajna-sala of Birla Temple. Pillars are inscribed with details of alchemical event. Photo credit: Dash

Bhagwan Dash made an interesting point at the beginning of his book which helps explain why the topic of alchemy would appear openly in India in the 20th century, let alone within the sacred walls of temples. "Whatever may be the opinion in European countries," Bhagwan Dash wrote, "in India, [alchemy] is not considered a myth." (Bhagwan Dash, 1986)

He wrote that, in descriptions of successful alchemy in India even before the 6th century B.C.E., certain processes were kept secret to prevent the knowledge from being abused by unscrupulous people. There are two plaques in the Birla Temple in New Delhi. Here are photographs of the inscriptions and Bhagwan Dash's translations, which I will present as images in order to more accurately preserve his work and the Sanskrit characters:

Inscription in one of the pillars in the Yajna-sala of the New Delhi Birla Temple. Photo credit: Dash

"In the month of *Caitra* (name of the month according to the Hindu calender corresponding to March-April) of the *Vikrama Saṃvat* 1999 (1942 A.D.), one Śrī Kṛṣṇa Lāla Śarmā, Rasa Vaidya Śāstrī, originally hailing from Pubjab came from Ṛṣikeśa to Delhi to demonstrate the practical method of preparing gold out of mercury. On this occasion, the secretary of Mahātmā Gāndhī, Śrī Mahādeva Desāī, Gosvāmī Gaṇeśa Datta and Śrī Jugala Kiśora Biralā (the noted industrialist of India) were present. In front of them, 200 *tolās* or 2½ seers (1 *tolā* is approximately 12 Grams) of mercury was mixed with one *tolā* of the powder of a drug (identity undisclosed) and the whole thing was kept over fire for half an hour. Thereafter, the mercury became gold. This process was repeated, and as such 18 seers of gold was prepared."

Translated inscription in one of the pillars in the Yajna-sala of the New Delhi Birla Temple. Text credit: Dash

पुनश्च

ज्येष्ठ शुक्ला ९ संवत् १९९८ तारीख २७ मई १९४१ ई. को बिरला हाउस नई दिल्ली में श्री पं. कृष्णापाल शर्मा ने हम लोगों के सामने एक तोला पारे से लगभग एक तोला सोना बनाया था। पारा एक सीटे के अन्दर डलवाया गया। उसमें एक जड़ी बूटी का सफेद चूर्ण और दूसरा पीला चूर्ण जो शायद ही वजन में एक या डेढ़ रत्ती होगा, पारे में डाले गये। फिर वह सीटा गीली मिट्टी से बंद कर दिया गया और गीली मिट्टी के दियों के संपुट में बंद करके आग पर रखवा दिया। लगभग पौन घंटे तक पंखों से आग को धोंक कर तेज कराया गया। जब कोयले जल कर राख होने लगे, तब उसे पानी में छुड़वाया। दियों के संपुट के अन्दर से सोना बनकर निकल आया। तोलने पर कोई एक-दो रत्ती कम एक तोला उतरा। बिलकुल खरा था। इस क्रिया के अन्दर क्या रहस्य था, वह दोनों चूर्ण क्या थे, यह हमें मालूम नहीं हो सका। पंडित कृष्णापाल इस सारी क्रिया के कराते समय हम से दस-पन्द्रह फुट की दूरी पर खड़े रहे। उस समय श्री अमृतलाल वि. ठक्कर (प्रधान मन्त्री अ. भा. हरिजन सेवक संघ) श्री गोस्वामी गणेशदत्त जी लाहौर, बिरला मिल दिल्ली के सेक्रेटरी श्री सीताराम जी खेमका, चीफ इंजीनियर श्री विल्सन और श्री वियोगी हरि उपस्थित थे। हम सबको यह क्रिया देखने का आश्चर्यं हुआ। श्रीमान सेठ जुगलकिशोर बिरला ने सारी क्रिया कृपाकर हम लोगों को दिखलाई।

मार्गशीर्ष कृष्णा ५- सं. २००० दिल्ली

हस्ताक्षर
१- अमृतलाल वि. ठक्कर
२- सीताराम खेमका
३- वियोगी हरि

स्वर्गीय पं. कृष्णपाल शर्मा ने सबैद्य झारजी को यह विद्या एक साधु ने बतलाई थी किन्तु उनकी दृष्टि में सुपात्र व्यक्ति न मिलने के कारण उन्होंने यह विद्या किसी को भी नहीं बताई।

Inscription in one of the pillars in the Yajna-sala of the New Delhi Birla Temple. Photo credit: Dash

18 • ALCHEMY, THE PRECURSOR TO MODERN SCIENCE

> "On the first day of *śukla pakṣa* (bright fortnight) in the month of *Jyeṣṭha* (name of a month according to Hindu calender corresponding to May-June) of Saṃvat 1998 i.e. 27th May, 1941, Pt. Kṛṣṇa Lāla Śarmā in our presence (names of these persons are given below) prepared one *tolā* of gold from out of one *tolā* of mercury in Birala House, New Delhi. The mercury was kept inside a fruit of *rīṭhā* (bot. *Sapindus trifoliatus* Linn.). Inside this, a white powder of some herb and a yellow powder which were perhaps one or one and half *rattī* (one *rattī* is equal to 125 mg.) in weight were added. Thereafter, the fruit of *rīṭhā* was smeared with mud and kept over fire for about 45 minutes. During that process, the fire was made stronger with the help of a fan. When the charcoal after ignition became ash, water was sprinkled over it. From inside the fruit which originally contained mercury, gold came out. In weight, the gold was 1 to 2 *rattis* less than one *tolā* (originally used). It was pure gold. We could not ascertain the nature as well as the identity of both the powders which were added to the mercury were not disclosed to us. During the whole experiment, Pt. Kṛṣṇa Lāla was standing about 10 to 15 ft. away from us (site of performance). During this time Śrī Amṛta Lāla V. Thakkara (Chief Secretary, *Akhil Bhāratīva Sevaka Saṅgha*), Śrī Gosvāmī Gaṇeśa Dattajī (of Lahore), Secretary of Birla Mill in Delhi Śrī Khemakā, Chief Engineer Mr. Wilson and Śrī Viyogī Hari were present. We were all surprised to witness this performance. Seth Śrīmān Jugala Kiśora Biralā was kind enough to show us this performance.
>
>   Signed :   (1) Amṛta Lāla V. Thakkara
>
>   (2) Sītā Rāma Khemakā
>
>   (3) Viyogī Hari
>
> *Mārgaśīrṣa Kṛṣṇa* 5, *Saṃvat* 2000, Delhi (This was perhaps the date of installation of the plaque).
>
> Late Pt. Kṛṣṇa Lāla Sarmā, Rasa Vaidya Śāstrī learnt this technique from a saint named Nārāyaṇa Svāmī. But in the absence of a suitable disciple, according to him, he did not teach this technique to any body."

Translated inscription in one of the pillars in the Yajna-sala of the New Delhi Birla Temple. Text credit: Dash

If true, the Birla Temple mercury-to-gold transmutation would be an amazing story. Certainly, there are enough names listed to allow a curious person in India to do a deeper investigation.

Decline of Alchemy

In the 1600s and 1700s, the scientific method began to take form, although some respected scientists of the day — for example, Isaac Newton and Robert Boyle — conducted alchemical experiments. Eventually, the scientific method took over. In learned circles, chemistry was favored over alchemy, and astronomy was favored over astrology. Atomic science, born anew in 1895, had no direct ancestor.

The contempt for alchemy by the literati was no trivial matter. A bitter article in 1913 by Leonard Keene Hirshberg, a physician, scientist and writer, in *Harper's Weekly*, shows just how much men and women of science despised alchemy:

> Time was when every third savant was a dignified alchemist, and every fourth one an eminent astrologer. Those were the old days, to be sure. ... The savants of that past era held it as a sane ambition of their hopes so to conquer the baser things of the physical world that they might at will convert cheap minerals into fine gold. They held to the principle that there was an underlying agent or elements in all matter with a real capacity for changing granite or tin into gold; chalk or coal into diamonds: and even death itself into life.
>
> As the great new sciences of chemistry and physics became firmly established, nearly every serious-minded investigator ceased his attempt to transmute iron, zinc and the baser metals into gold or silver. The men who so continued to experiment were sent to [mental hospitals]. In brief, they were ... called madmen, and dubbed liars or fools. (Hirshberg, 1913, 21)

After alchemy's heyday in medieval times and the next few centuries, John Dalton (1766-1844), an English chemist, meteorologist and physicist, published a theory in 1808 that reinforced the scientific view that transmutation was impossible. In doing so, Dalton provided support to his scientific colleagues, who were struggling to distinguish their reputations from their brethren who chose to walk down the alchemical path.

In his seminal work, *New System of Chemical Philosophy*, Dalton proposed that atoms were the smallest possible units and that nothing could change within them. He also said that atoms were indivisible and, therefore, transmutations of one element to another were impossible. Dalton's treatise was broadly accepted by the scientific community at the time, and it kept alchemy at bay for almost 100 years.

In 1880, the scientific world still had the upper hand and managed to keep the dreaded alchemy out of its hallowed halls. Professor H. Carrington Bolton of Trinity College in Hartford, Connecticut, gave a lecture on alchemy at the New York Academy of Sciences on Nov. 15, 1880. Alchemy had not yet made its revival, and Bolton spoke about it calmly — for now. (*New York Times*, 1880, 2)

CHAPTER 3

A Strange Glow; An Unusual Shadow

Photography Opens the Door to the Discovery of Radiation (1896)

In the late 1880s, John Dalton's 1808 model of the indivisible and immutable atom still served as the prevailing view of the nature of elements. Dalton's model had helped to distinguish legitimate science from pseudoscience. Alchemy had been cast aside.

But this separation was not to last. Scientists were about to learn that, in fact, the atom was divisible and it comprised subatomic particles, giving rise to the new field of atomic research. Within a decade of the newer understanding of the atom, this newly discovered mutability of the atom inspired renewed interest in alchemy.

Discovery of X-Rays

The first event that opened the doors for atomic science and paved the way for the new concept of the atom took place in 1895. Wilhelm Roentgen, a German physicist (1845-1923), accidentally discovered a new kind of radiation, which he called X-rays.

Other scientists had observed darkened photographic plates and other effects from Crookes tubes during the previous 20 years, but Roentgen was the first person to systematically study and name X-rays.

Roentgen was studying the effect of fluorescence, working with glass tubes coated with a special material on the inside. As he passed an electric current through the inside of the tube, streams of electrons hit the inside wall of the coated glass and, as expected, caused it to glow.

Roentgen had no reason to think that he could see the fluorescent glow occur on surfaces outside the glass tube because electrons did not have the energy to go through glass. To his great surprise, one day in the darkened room, he noticed that a coated screen sitting on a bench a few feet away began to glow during the experiment. He also knew that the effect could not be from visible light because he had fully covered the glass tube. He began a series of tests to see whether he could block the unknown rays, first by placing thick cardboard and, later, other materials between the glass tube and the screen. To his amazement, he continued to see the screen glow.

 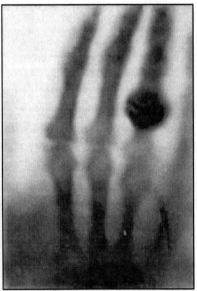

Wilhelm Roentgen and one of his first X-rays, taken of his wife's left hand on December 22, 1895, with a ring on one finger.
X-ray photo Credit: National Library of Medicine

For several days, he ate and slept in his lab while trying all sorts of variations, doing his best to be sure that he was not making an error. He placed the screen behind a thousand-page book, yet it continued to glow fluorescent. He put a variety of objects that were readily available to him in the way: packs of cards, pieces of wood, sheets of metal. The unknown rays went through all of them. However, the thicker materials partially diminished the brilliance of the light.

Eventually, he placed sheets of leaded glass in the way and noticed that they drastically reduced the effect of the mysterious rays. Thus, he began to understand that the rays could penetrate many materials but not lead or other dense materials. At one point, while he was holding up an object between the tube and the screen, he saw a ghostly image of the bones of his hand. Imagine his shock! The screen allowed him to see what was happening in real time, and he found that photographic plates could permanently record the shadows created by the transmitted X-rays. The first medical X-ray photograph was of his wife's hand. Within months, the first radiology departments were established in hospitals.

Discovery of Radioactivity

Word traveled fast. In 1896, Antoine Henri Becquerel, a French physicist (1852-1908) inspired by Roentgen's work with photographic plates, worked on a new line of experiments. He thought that, if he placed certain phosphorescent material, like uranium salts, on photographic plates and exposed them to sunlight, he might see evidence of Roentgen's X-rays. At first, he did see some impressions on the photographic film, and this was interesting but not profound.

The next day, he prepared another set of films, but the sky was cloudy. So he left the salts sitting on the film in a drawer, in the dark, without expecting to see any unusual results. Because the sun didn't come out for another two days, he decided to develop the film, anyway. To his amazement, he saw a clear image of the objects and realized that the sun had nothing to do with whatever mysterious forces had made the image; some other rays were responsible. This spontaneous emission of these mysterious rays marked the commonly accepted version of the discovery of radioactivity. (Becquerel, 1896)

 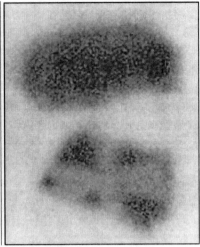

Henri Becquerel; Image of his photographic plate after he exposed it to uranium salt. The lower-right image shows the shadow of a piece of metal, a Maltese Cross, which he placed between the plate and the uranium salt.

Before Radioactivity

Thirty-eight years earlier, Abel Niépce de Saint-Victor (1805-1870), a French photographic inventor, had observed a similar phenomenon while using metal salts to produce color photography. Niépce reported his findings to the French Academy of Science in 1858. Editor Michael Ravnitzky found this history while reviewing a draft of this book.

A 2001 journal article published by Fathi Habashi credits Niépce rather than Becquerel with the discovery of radioactivity. (Habashi, 2001) Philip Gibbs, the founder of the online scientific preprint server viXra, an alternative to the better-known arXiv, shed light on this history in a blog article. Here is an excerpt:

> Like Becquerel, [Niépce] was studying the effects of light on various chemicals and was using photographic plates to test the reaction. He also used uranium salts and found that they continued to blacken the plates long after any exposure to light had been stopped. Fluorescence and phosphorescence had been known for many years, and

Niépce knew that this new observation did not conform to either phenomenon. He reported his results to the Academy of Sciences in France several times.

A few scientists, including Foucault, commented on the findings, but no one had a good explanation. Surprisingly, no one seems to have tried to replicate them, and it is likely that everyone thought that there was most likely some experimental error. In any case, Niépce and his discovery were soon forgotten.

When Becquerel rediscovered the same result as Niépce, the situation was very different. By then X-rays were known, and physicists were ready to appreciate that another new type of ray could exist. One physicist, Gustave Le Bon, pointed to the prior work of Niépce, but Le Bon was ridiculed. Any further chance that Niépce might gain some recognition was extinguished when the Nobel committee awarded the physics prize to Becquerel. (Gibbs, 2010)

Tiiu Ojasoo, a biochemist who studied at the Pierre and Marie Curie University, also analyzed this history. He offered an additional insight:

> According to Le Bon, therefore, the person who made the first observation [of radioactivity] was the unpretentious Niépce de Saint-Victor, who was disdained during his lifetime and forgotten after his death. Niépce rightly claimed that uranium salts could emit radiations that mark photographic plates in the dark but erroneously considered these radiations a form of captured light, of phosphorescence. His experimental evidence was neglected, but his misconception of the nature of the phenomenon was upheld by the general scientific community. (Ojasoo, 1996)

The final twist to this story is that Niépce was an acquaintance of Alexandre Edmond Becquerel, Henri Becquerel's father, who knew of Niépce's observation. Alexandre Edmond Becquerel described Niépce's work in a book he wrote in 1868. (Becquerel, 1868, 50)

Naming Radioactivity

In 1898, Maria Sklodowska Curie, a Polish-born and naturalized French physicist and chemist (1867-1934), after examining Becquerel's rays, coined the term radioactivity.

This was still several years before scientists got a clear understanding of the structure of the atom, or its major subatomic particles: the proton, the neutron and the electron.

By 1896, scientists were aware of the myriad chemical elements. They believed that atomic elements were not only primary constituents of matter but also indivisible and immutable. The concepts of nuclear fission, nuclear fusion, nuclear transmutation, radioactive decay, and nuclear energy were still many years away.

Joseph John Thomson, a British physicist (1856-1940), discovered the electron in 1897. The electron, with its negative charge, was the first of the three primary subatomic particles to be discovered. Thomson played a part in shaking up the prevailing view of atomic theory; it was the first time that anyone suggested that atoms were composed of divisible parts. The proton and neutron were yet to be recognized. (Thomson, 1897)

Alchemy Returns

The birth of atomic science was in full swing and, with it, came a resurgence in alchemy, particularly in France. An Oct. 17, 1897, article in the *New York Times*, "The Revival of Alchemy," reported on another lecture by Bolton, given at a New York meeting of the American Chemical Society. The full essay was published in December that year in *Science*. Bolton explained that alchemy was on the rise, but this time his tone, unlike at his lecture in 1880, was angry:

> Fraud, folly and failure have been deeply written into the annals of alchemy in all ages; it was early characterized as an 'art without art, beginning with deceit, continued by labor and ending in poverty,' and in modern times, its

extravagant pretensions have been condemned by an exact and critical science, yet notwithstanding there are today indications of a resuscitation of the captivating theories and of renewed attempts at their practical application, of great interest to students of the intellectual vagaries of mankind.

Belief in the possibility of prolonging life by an artificial elixir and of transmuting base metals into silver and gold was generally entertained in the Middle Ages, not only by the ignorant masses but even by serious-minded philosophers imbued with all the learning of the time; and the popular faith was sustained by the tricks of unprincipled impostors who found it profitable to prey upon the credulity and avarice of their fellow men. (*New York Times*, 1897, 6; also *Science*, 1897, 853)

After expressing his contempt for alchemists, Bolton discussed two events that occurred in the United States, one of which, he wrote, "seems to threaten financial revolution."

Bolton said that Edward C. Brice of Chicago claimed he could manufacture gold from pure antimony. The other event that worried Bolton was Stephen H. Emmens' claim that he had discovered a new material that was a hybrid of silver and gold and that he had the power to change it into pure gold. According to Bolton, Emmens sold $954 of this material to the United States Assay Office. (Bolton, *Science*, 1897, 862; *Morning Times*, 1897) There was no apparent impact on the financial sector.

Bolton's article in *Science* contained a wondrous depiction of the life of an alchemist. Bolton described the trappings of wizardry in fiction, such as the ability to project psychic forces, the donning of ceremonial robes, and the discipline of cultivating certain mental states.

Atomic Science Advances

Roentgen's X-rays and Becquerel's radiation laid the groundwork for the research of Marie Curie, and her husband, Pierre Curie, a French physicist (1859-1906). Marie studied Becquerel's rays produced by the uranium salts and began measuring their intensity. She and Pierre discovered polonium and radium in 1898. These elements were naturally radioactive, and radium soon became a favorite substance of researchers who were exploring the new atomic world.

The same year, two men who would later earn Nobel prizes each made significant contributions to science. Their roles in the next two decades exemplified the coming split between chemistry and physics.

In 1894, Sir William Ramsay, a Scottish chemist (1852-1916), along with English physicist John William Strutt (1842-1919) (the father of Robert John Strutt), discovered argon, and in 1895, Ramsay discovered the presence of helium on the earth. Helium had been identified on the surface of the sun in 1868.

Ramsay discovered the noble gases krypton, neon and xenon in 1898. They were called "noble" because scientists thought they would not chemically interact with any other element, although now it is known that they may interact under certain exceptional conditions.

In 1898, Sir Ernest Rutherford, a New Zealand-born British physicist (1871-1937) who worked extensively in the United Kingdom, followed Becquerel's work with uranium and pursued the mystery of the radiation emitted from it.

Rutherford separated "Becquerel's rays" into two types, which were later determined to be particles (with mass), not rays (without mass). Rutherford named them "alpha" and "beta" rays.

In 1900, Paul Villard, a French chemist and physicist (1860-1934), observed a third form of radioactive emission coming from uranium that was more penetrating than alpha or beta particles. In 1901, Rutherford christened it "gamma radiation."

In 1901, Rutherford, along with Frederick Soddy, an English radiochemist, (1877-1956), went to work to solve the mystery of an unexplained emanation coming from a naturally radioactive element. This collaboration between a physicist and a chemist would make history. Before going further, it would be helpful to provide a review of what's inside the atom.

CHAPTER 4

Atomic Boot Camp

Basic Training in Nuclear Science for the Rest of Us

For nearly a century, scientists have known that all matter is composed of elements, elements are composed of atoms, and atoms are composed of three primary subatomic particles: the proton, neutron and electron.

Protons have a positive electrical charge, electrons have a negative charge, and neutrons have no charge. Protons and neutrons sit close together in the center of the atom and compose the nucleus. (See diagram of Carbon-12) Electrons, which have a negative electrical charge, remain outside of and orbit the nucleus.

Different Elements

All matter exists in the form of specific elements — for example, hydrogen, oxygen, and carbon. Each element is distinguished by the number of protons in its nucleus. At present, 98 elements are known to exist in nature, and a few others have been synthesized in laboratories.

When the number of protons inside the nucleus increases or decreases, the atom changes from one kind of element to another. For example, a proton added to a nitrogen atom changes it to an oxygen atom. This is called a nuclear transmutation. The first diagram below shows, on the left, one of the simplest atoms, a form of hydrogen called deuterium. Normal hydrogen has only one proton in its nucleus, but this variety of hydrogen, called deuterium, has a neutron, as well. On the right, the diagram shows a nucleus with two protons; this is the element helium. In this case, it's a variety of helium called helium-3, with has three particles in its nucleus: two protons and one neutron.

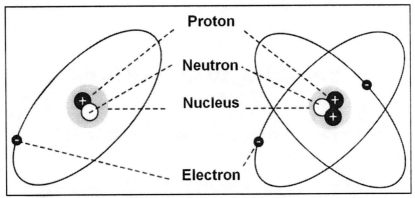

Deuterium atom (left): one proton and one neutron in the nucleus. Helium-3 atom (right): two protons and one neutron in the nucleus.

Different Isotopes

Most elements exist in a variety of forms. Just as chocolate comes in different varieties, so do elements. However, different varieties of chocolate are still chocolate. A variation of an element is called an isotope. Each isotope is slightly different from other varieties of the same element. The difference between isotopes is that they have different numbers of neutrons in each nucleus, but the number of protons in each nucleus stays the same. An isotope of helium is still helium; an isotope of hydrogen is still hydrogen.

The pair of diagrams below provides an example. The one on the left shows a variety of helium called helium-3. It has three particles inside its nucleus: two protons and one neutron. The one on the right shows helium-4, which has two protons and two neutrons in its nucleus. Both are different isotopes of the same element, helium.

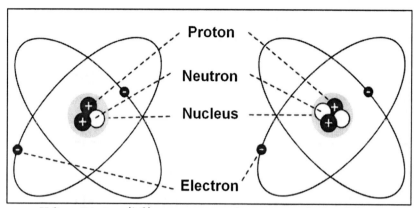

Helium-3 isotope (left): two protons and one neutron in the nucleus.
Helium-4 isotope (right): two protons and two neutrons in the nucleus.

Nearly all elements have a variety of isotopes. Some elements have many isotopes; some have only a few. Some isotopes of a given element are more abundant than the other isotopes. For example, a lump of coal is mostly carbon. Isotopic analysis of the carbon reveals that most of it exists as the carbon-12 isotope.

Carbon-12 has six protons and six neutrons in its nucleus. Carbon has a total of 15 isotopes. Stable carbon-12 usually makes up 98.93% of the total amount of any sample of carbon. One of the other isotopes, stable carbon-13, for example, makes up only 1.07% of naturally occurring carbon.

The ratio between carbon-12 and carbon-13 is normally very nearly the same, whether it is measured in Colorado or Kiev. This phenomenon applies to all elements, not just carbon. The percentage of each isotopic abundance does not usually vary from its natural state. Because of this, these ratios act like scientific fingerprints. If scientists find a sample of an element that contains abnormal isotopic ratios, they know that an unusual event has occurred.

Generally, there are three types of events. First, environmental and biological factors can segregate some of the isotopes and cause minor shifts in the ratios between isotopes at certain locations over long periods, and forensic scientists can use this data to correlate biological samples to specific geographical locations. Second, a wide variety of man-made processes can be used to separate and concentrate isotopes.

Methods include diffusion mechanisms, centrifuges, electromagnets, or lasers. Isotopic separation is the way low-grade uranium is enriched to make high-grade uranium. Third, nuclear reactions can cause isotopic shifts that add or remove neutrons from isotopes. Although the first two type of events cause isotopic fractionation, only nuclear reactions make direct changes among the particles in the nuclei.

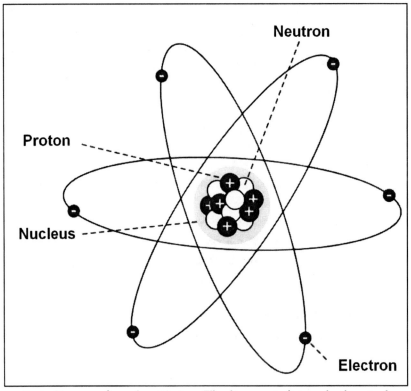

Basic diagram of a carbon-12 atom. The diagram isn't to scale; the actual distance between the orbiting electrons and the nucleus is much greater than shown here. Also, the specific arrangement of electrons in their valence shells is not depicted here.

A Matter of Power

Elements change, or transmute, into other elements by the addition or subtraction of protons. Isotopes change into other isotopes by the addition or subtraction of neutrons. Both kinds of changes are nuclear, not chemical. Chemical changes involve either the addition or subtraction of electrons of a single atom, or the regrouping of atoms among themselves; they do not involve changes inside the nucleus. Not all reactions that occur in nature or science are equal. There are big differences between reactions that can cause a chemical change and those that can cause a nuclear reaction. One of the biggest differences is the amount of energy required to initiate a reaction. A nuclear reaction typically requires one thousand to one million times more energy than a chemical change.

Primarily, two fundamental physics forces affect protons: the electromagnetic force and the strong force. Neither relinquishes its power and control over protons without a fight.

These forces, the electromagnetic and the strong, prevent protons from jumping from one atom to another and help provide stability for matter.

Under most circumstances, the electromagnetic force repels protons from each other like the north poles of magnets repel each other. It is no easier to squeeze two protons together than it is to try to press the north poles of two magnets together with your bare hands and expect them to stick. With protons, the exception is that, if they are squeezed incredibly close to each other, the strong force overcomes the electromagnetic force, which then slams the protons together. The strong force works only at very short distances. Conversely, to separate protons from each other, an immense force is required to free protons from the stranglehold of the strong force.

But no chemical process ordinarily has enough energy to bring protons together or to separate them. For a century, scientists have known that the sheer muscle required to make these kinds of changes can be triggered only by high-energy physics. This typically takes place through the use of particles that are emitted with high levels of energy. A moving projectile such as a bullet may be small, but the rate at which it is traveling gives it enough power to shatter dense material.

At the turn of the last century, chemists and physicists began to realize that radioactive materials such as radium had the unusual characteristic of emitting particles at high energies. As scientists

Frederick Soddy and Ernest Rutherford began studying these unusual occurrences, the two made one of the most profound discoveries in modern science.

CHAPTER 5

Don't Call It Transmutation!

Soddy Envisioned It, Rutherford Avoided It: the First Observation of Natural Nuclear Disintegration (1902)

Let's review the development of atomic science thus far: Roentgen had discovered X-rays in 1895, Becquerel had rediscovered radiation in 1896, the Curies had discovered polonium and radium in 1898, and that year, Rutherford had separated "Becquerel's rays" into two types and named them "alpha" and "beta."

Chemist Wanted

The same year, when Rutherford was 27, he left the Cavendish lab at the University of Cambridge for a faculty position in the Physics Department at McGill University in Canada. He and other people began experimenting with another radioactive element, called thorium. Early in 1900, Rutherford had reached an impasse. He knew that something was emanating from the thorium, he knew he was dealing with a chemical problem, but as a physicist, he didn't have the knowledge and techniques to figure out what it was. He needed the help of a chemist. In October 1901, at Rutherford's request, Frederick Soddy, 24, began to work with him on the problem. Soddy had left Oxford University the year before and had joined McGill as a demonstrator in the Chemistry Department.

Ernest Rutherford, 1908 Photo

Frederick Soddy, 1905 Photo

Soddy and Rutherford figured out that the emanation was distinct from thorium. They didn't say what that emanation was; the precise identification of it would come later when Soddy went to work with William Ramsay.

Walking on Eggshells

Ever so cautiously, they mentioned in their published papers that the emanation behaved like an argon gas. But they did not state that they observed the transmutation of one element to another, or, as Rutherford feared, their peers would have their heads as alchemists. Instead, they went to great lengths to downplay their achievement.

Rather than report their finding as a transmutation from one element to another, they said that thorium was changing to an unidentified thorium-like material, which they called thorium-X. They did not claim to have transmuted elements, and they avoided using the word transmutation. Rutherford was well aware of and concerned about the risk of being associated with alchemy. Instead, he and Soddy called it disintegration.

The major significance of their work was that they figured out that radiation was linked to atomic changes and that atomic changes could, in fact, take place. This contradicted 100 years of belief by scientists that atomic changes could not take place on Earth.

The working relationship between Rutherford and Soddy marks one of the most effective collaborations of a chemist and physicist in science history. Each man approached the scientific mystery and the uncomfortable allusion to alchemy, and responded to their findings in markedly unique ways.

The following list distinguishes the concepts of disintegration and transmutation:

Disintegration Versus Transmutation*

Case 1: Natural Nuclear Disintegration
An *unidentified* nuclear change that occurs spontaneously in an unstable nuclide. Example: Rutherford and Soddy's 1902 observation of the disintegration of thorium into "thorium-X."

Case 2: Natural Nuclear Transmutation
An *identified* nuclear change, from one nuclide to another, that occurs spontaneously in an unstable nuclide. Example: Ramsay and Soddy's 1903 observation of the transmutation of thorium into helium.

Case 3: Man-Made Nuclear Transmutations
An identified nuclear change, from one nuclide to another, that occurs at will in a stable nuclide. Example: Collie and Patterson's 1913 transmutation of hydrogen into helium and neon. (Chapter 13)

> * This list uses the terms nuclear and nuclides, but those terms were not used until after Rutherford confirmed the existence of the nucleus in 1913.

These three cases help to distinguish the type of observation and the nature of the claims made by the researchers. In Cases 1 and 2, the same type of process occurs, but in Case 2, the researchers went further than the researchers in Case 1 and distinguished the changed nuclides. This is significant for the knowledge attained as well as for the shift in scientific attitude of such a controversial topic.

Cases 1 and 2 rely on the spontaneous changes of radioactive elements. In Case 3, researchers went further again and made changes to nonradioactive elements, further advancing understanding of the new science.

Alchemy Revisited

Author and historian Richard E. Sclove, after immersing himself in the Oxford University archives, published an extensive essay on Soddy's history. He found that, before Soddy and Rutherford made their discovery, Soddy shared the same contempt as his fellow academicians for the scandalous forays of alchemy.

But Soddy wasn't completely convinced of the mainstream view; he saw that the more recent developments in science suggested an inconsistency. Soddy knew that any legitimate scientist at the turn of the century who suggested that an element could change to another element was risking his or her reputation. In an admirable example of following the scientific process, he followed the data and began to reconsider his views. That required no small measure of courage and confidence, particularly for a young man just beginning his career in science.

Instead of overlooking the inconsistency, Sclove wrote, Soddy went deeper into the historical research. The result was a short, unpublished paper called "Alchemy and Chemistry," in which he reconceived his view of chemical history. Rather than see alchemy as an illegitimate offshoot of chemistry, Soddy began to see that chemistry originated from alchemy. Soddy, however, still held a hard line against the fraudulent medieval claims of transmutation. (Sclove, 1989, 166)

Rutherford was obsessed with the particle physics and the characterization of the radiation. It was Soddy, however, who had the specific intent and, in fact, a burning desire to accomplish elemental transmutation. Sometime in the months preceding their moment of discovery in the fall of 1901, Soddy wrote his manifesto.

"The constitution of matter," Soddy wrote, "is the province of chemistry; and little indeed can be known of this constitution until transmutation is accomplished. This is today, as it has always been, the real goal of the chemist before this is a science that will satisfy the mind." (Sclove, 1989, 167)

In 1901, scientists knew that elemental changes took place in the fiery uninhabitable realms of the stars. But to think that such radical transformations could take place in the tepid domain of Earth was more than a bold step, as Sclove explained.

> Soddy's idea of trying to transmute the elements was both imaginative and audacious for it was no small step from accepting the possibility that transmutation might occur under the so-called transcendental conditions present in stars (Lockyer's view), to imagining that somehow the same

processes could be made to occur on Earth. Moreover, insofar as Soddy had previously identified himself with the chemical tradition, in setting a long-range goal for chemistry, he was presumably setting a goal for himself. (Sclove, 1989, 167)

Eureka Moment

Soddy described his moment of discovery to Muriel Howorth, who spent many long hours with him, gathering his stories and recording his memories, and wrote his biography.

> "I simply inserted into the air-stream," Soddy said, "between the thorium compound and the ionization chamber in succession, a number of powerful chemical reagents such as magnesium powder, platinum black, lead chromate, and zinc dust at temperatures up to white heat — one or other of which would have absorbed every known gas before the discovery of argon. The emanation passed through all of them completely, unabsorbed and unaffected.
>
> "Eight years earlier, this inability to absorb the gas would have conveyed nothing whatsoever, even to a chemist, but in 1900, I knew that it meant that thorium was changing spontaneously into an argon gas." (Howorth, 1958, 82)

Soddy knew it was one of the noble gases like argon, which was discovered by William Ramsay, because it wouldn't interact with the other elements he tried to put in its way.

When Howorth heard this, she understood the significance of the discovery. "I knew those words were world-shattering: the result of that one experiment had a colossal portent," she wrote. "It conveyed the tremendous and inevitable conclusion that the element thorium was slowly and spontaneously disintegrating itself. Nature had proved to be the great alchemist transmuting one element into another. She had shown Man the way: undoubtedly now he could follow up the trail."

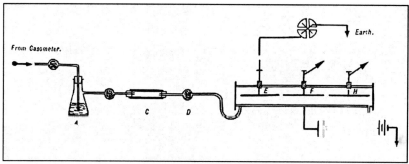

Rutherford-Soddy experimental schematic. A current of air passes through various filters (A) until it reaches the target powder, resting inside the glass cylinder (C). The air picks up and carries the emanation from the powder through another filter (D) and then into a brass cylinder in which the emanation is detected by three electrodes for analysis (E,F,H).

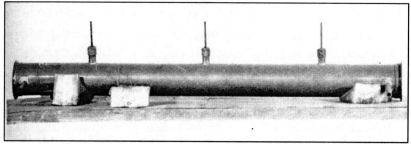

Rutherford-Soddy experiment: Photograph of the brass cylinder.

Responding to Howorth's interest, Soddy explained what happened when he fully realized what he and Rutherford had done:

"My mind was always occupied with transmutation," Soddy said. "That is natural; I was a chemist. You will remember my paper on 'Alchemy and Chemistry'; I made that goal quite clear. Also at that time, I had been working on the lectures on gas analysis which I had been asked to give in the University.

"That is why, perhaps, when Rutherford showed me the emanation which was not thorium, nor alpha nor beta particles, but [something] which could be blown about, I drew his attention to the fact that it would be a gas. I do want to show, as fairly as I can, that it was not very strange that I should be the first to discover and announce quite confidently the natural transmutation of radioactive elements. After all, I had tried the emanation out with every reagent that should have absorbed any known gas, and it had passed right through each one. It was natural to infer that it must be one of the newly found argon gases, though not one that [William] Ramsay had already discovered.

"I was, of course, tremendously elated to have discovered transmutation — the goal of every chemist of every age, but looking back through the years it was, perhaps, the courage to express my conviction which was the laudable feature about it. I had written in my paper, 'Alchemy and Chemistry,' that transmutation had not yet been found; yet when the time came to investigate the phenomenon, the whole thing seemed too devastatingly simple. The fact that this was, in reality, transmutation, flashed through my brain, and I could hardly believe what I knew to be true.

"I remember quite well standing there transfixed as though stunned by the colossal import of the thing and blurting out — or so it seemed at the time: 'Rutherford, this is transmutation: the thorium is disintegrating and transmuting itself into an argon gas.' The words seemed to flash through me as if from some outside source.

"Rutherford shouted to me, in his breezy manner, 'For Mike's sake, Soddy, don't call it *transmutation*. They'll have our heads off as alchemists. You know what they are.' After which, he went waltzing round the laboratory, his huge voice booming 'Onward Christian so-ho-hojers.'

"I realized that if the thorium emanation was transmuting itself into another element — an argon gas — no doubt other elements were undergoing a process of natural transmutation in much the same way. This was the start of the disintegration theory of radioactive substances." (Howorth, 1958, 82-83)

Disintegration Versus Transmutation

Rutherford and Soddy took pains to avoid the association with transmutation. The word does not appear in their papers. Soddy, at least, knew they had done more than just disintegrate thorium. He knew they had transmuted thorium into an argon-type gas. Soddy argued with Rutherford that they had transmuted elements (see below) so it is unclear whether Rutherford lacked Soddy's confidence in the data or whether he lacked Soddy's courage to confront an obsolete scientific belief.

The first of Rutherford and Soddy's series of papers simply announced their discovery of thorium-X and said that they had identified the radioactive emanation as a chemically inert gas. The second paper was bolder because they suggested that they had made chemical changes.

A letter written by Rutherford when he submitted the paper for publication reveals his thoughts. The letter was reproduced by science historian Lawrence Badash. At the same time that Rutherford submitted the second paper to the Chemical Society (later known as the Royal Chemical Society), he sent copies to Sir William Crookes, a prominent chemist who was editor and publisher of *Chemical News*, to solicit his help and facilitate acceptance of their paper. I have marked a key sentence in his letter in italics. It is the only instance I have found, although private, in which Rutherford uses the word transmutation.

> Dear Sir William, I have to thank you for the reprints of your two Royal Society papers which I received yesterday and in which I was very interested. The blue mercury button in the hydrogen tube is a most extraordinary manifestation and shows how much is still to be investigated in that field.
>
> I am forwarding to you some reprints of my previous papers on Radioactivity. I am sending you by this mail a [manuscript] by Mr. Soddy and myself on the "Radioactivity of Thorium" which we are forwarding at the same time to the Chemical Society. We have found that the Th-X

[thorium-X] like Ur-X [uranium-X] loses its radioactivity in a G.P. [geometric progression] with the time while the deactivized thorium regains its activity with time. I think we have conclusively shown that most of the radioactivity is due to a production of Th-X at a uniform rate by the thorium and that this Th-X decays with time. An equilibrium point is reached when the rate of production is balanced by the rate of decay. We have strong evidence that uranium and radium behave similarly, only that the time rate of change is different. All these processes are independent of chemical and physical conditions, and we are driven to the conclusion that the whole process is sub-atomic.

Although of course it is not advisable to put the case too bluntly to a chemical society, I believe that in the radioactive elements we have a process of disintegration or transmutation steadily going on which is the source of the energy dissipated in radioactivity. [emphasis added]

But if you have time to glance over the paper, I think you will see the evidence is all experimental and that the obvious deductions from the facts are included in the last section. I am afraid that I have already troubled you sufficiently with my views, but Mr. Soddy and myself would both be obliged if you could do anything to facilitate the publication of the paper if difficulties arise over "atomic" views. Yours sincerely, E. Rutherford. (Badash, 1966, 89)

Due Credit

Soddy's realization, based on his work with Rutherford, formed the basis of his Disintegration Theory though Rutherford alone got the Nobel Prize for the theory. (More on that in Chapter 8)

"I know there are many [people] who credit Rutherford with the discovery of natural transmutation," Soddy said. "The Nobel Prize was no doubt awarded on that account. As I have always said, in working in a team it is invidious to take credit to oneself. I always try not to do so. I only want to show you how our brains were working, mine on transmutation and gases, Rutherford's on thorium and alpha ray emissions. (Howorth, 1958, 84)

Soddy was adamant about his role in the discovery, as he told Howorth many years later.

"Those who do not know the torment of the unknown cannot express the joy of discovery," Soddy said. "When I had finished my first experiment and had realized the immensity of the incredible discovery which I had made, the laboratory, before seemingly dark and befogged, like my mental vision — which had no inkling of the result I was about to obtain — suddenly seemed to open up before me, and a whole new glorious world swam into my sight."

"Rutherford did not — and you may rest assured of that — did not and could not have interpreted the results of the experiment," Soddy said, "but he had introduced me to the subject, and I have always given him credit for this." (Howorth, 1958, 85-86)

Howorth found a Feb. 22, 1936, letter that Soddy wrote to William Albert Noyes, the head of the Chemistry Department at the University of Illinois, that spoke directly to this point.

"I can absolutely deny that Rutherford had," Soddy wrote, "at that time, grasped the idea of atomic disintegration. ... After our six joint papers were published, I remember Rutherford taking me to task because people were saying that what we were saying was tantamount to 'transmutation,' and I had to convince him that it was transmutation and put him *au fait* with the chemical evidence to confute anyone who disputed it." (Howorth, 1958, 86-87)

In 1936, Noyes was one of the first historians who began omitting Soddy, in favor of Rutherford, for his contribution to the discovery of atomic disintegration. Noyes later apologized to Soddy and offered an excuse. "I made no attempt." Noyes wrote, "to write a complete or accurate history. That would have been impossible in so brief a paper."

History has left Soddy behind, and a deeper examination may reveal why. "These letters," Howorth wrote, "do give some indication of the resentment that he perhaps, not unnaturally, felt towards the end of his life against the injustice of having his part in the discoveries in McGill belittled, and, as time went by, seeing his name omitted altogether."

Both men were brilliant and energetic, and on the laurels of their work together, Rutherford rose quickly to great international fame. Rutherford's Nobel Prize in *chemistry* was awarded in 1908 "for his investigations into the disintegration of the elements, and the chemistry of radioactive substances." He did not share the prize with Soddy, the chemist who made the initial discovery. Journalist Richard Reeves wrote a book about Rutherford and had some additional insights about Rutherford:

> The prize was out of the ordinary in more ways than one, at least for Rutherford. He had always been rather single-mindedly dedicated to the advancement of physics, sometimes mocking chemistry and chemists. "All science is either physics or stamp collecting," he liked to say. (Reeves, 2008, 70)

The next step along the path of elemental transmutation was to definitively identify a transmuted element. In their last experiments together, Rutherford and Soddy investigated the emanation coming from radium. They speculated in an April 1903 paper that it might be helium, but they didn't get the experimental proof.

Soon after their radium experiments, Soddy left Rutherford and went to work with Ramsay. They nailed it: direct proof of the transmutation from one element to another. In the meantime, the Curies discovered the roots of modern nuclear energy.

CHAPTER 6

The Roots of Nuclear Energy

Mysterious Radium in Marie and Pierre Curie's Lab Burns Brightly but Is Not Consumed (1903-1904)

As sources of energy, coal, wood, cow dung and uranium have one thing in common: They all are sources of heat when burned. In the lowest forms of technology, cave dwellers burned wood for warmth. As technology progressed, people burned some of the same materials to produce steam, and the steam provided more options to use that energy. Eventually, steam was used to turn turbines to make electricity, and that form of energy was even more practical.

Fundamentally, nuclear energy uses uranium, and sometimes other materials, to create reactions that generate heat. The credit for the realization that radioactive materials possessed this powerful capacity to release heat goes to three scientists in Paris, France.

Marie Curie was fascinated with her and her colleagues' discovery of the element radium in 1898. She noticed that it always seemed to give off not just light but also heat. It had a hidden, internal source of energy. The study of radioactivity was only a few years old; the understanding of nuclear energy was a long way off.

Fire — that is, all forms of chemical combustion — had been used as a source of energy for nearly a millennium. The heat that Marie, her husband, Pierre, and their colleague, Gustave Bémont, discovered was the fundamental aspect of atomic science that, decades later, led to the use of an alternative to combustion: nuclear power.

Marie Sklodowska Curie, Pierre Curie

Powerful Heat

Unlike all chemical reactions, the radium never appeared to combust or disintegrate. Uranium, like any radioactive element, behaved the same way, but the level of radioactivity from radium was a million times stronger than that of uranium and thus more obvious to the Curies. Neither Marie Curie nor anybody else at that time had seen anything like it, as Marie Curie wrote in her memoir about Pierre:

> More striking still was the discovery of the discharge of heat from radium. Without any alteration of appearance, this substance releases each hour a quantity of heat sufficient to melt its own weight of ice. When well-protected against this external loss, radium heats itself. Its temperature can rise 10 degrees or more above that of the surrounding atmosphere. This defied all contemporary scientific experience. (Curie, 1963, 56)

Pierre Curie explains and demonstrates the luminous glow of highly refined radium at the Sorbonne. Drawing by André Castaigne

Inexplicable by Known Chemistry

In March 1903, Pierre Curie and Albert Laborde published the seminal paper reporting that radium was always giving off heat. Marie Curie wrote more about this mysterious heat in 1904:

> The properties of radium are extremely curious. This body emits, with great intensity, all of the different rays that are produced in a vacuum-tube. The radiation, measured by means of an electroscope, is at least a million times more powerful than that from an equal quantity of uranium.
>
> Radium possesses the remarkable property of liberating heat spontaneously and continuously. A solid salt of radium develops a quantity of heat such that for each gram of radium contained in the salt there is an emission of one hundred calories per hour. ... When we reflect that radium acts in this manner continuously, we are amazed at the amount of heat produced, for it can be explained by no known chemical reaction. The radium remains apparently

unchanged. If, then, we assume that it undergoes a transformation, we must therefore conclude that the change is extremely slow; in an hour it is impossible to detect a change by any known methods. (Curie, 1904, 461-6)

Atomic Transformation

Marie Curie's assumption about a transformation was supported by Rutherford and Soddy's reported disintegration of thorium the previous year. With heat released from the radium, she either assumed or intuited a direct relationship between the release of energy and the transformation of matter. She saw how her observation of heat was related to her colleagues' disintegration and transmutation experiments. One year later, Albert Einstein would publish his theory of special relativity and explain this transformation, expressed as $E=mc^2$. Curie also realized that her observations gave yet one more piece of direct evidence to disprove Dalton's theory of an indivisible atom.

> The emission of heat ... makes it seem probable that a chemical reaction is taking place in the radium. But this can be no ordinary chemical reaction, affecting the combination of atoms in the molecule. No chemical reaction can explain the emission of heat due to radium. Furthermore, radioactivity is a property of the atom of radium; if, then, it is due to a transformation, this transformation must take place in the atom itself. Consequently, from this point of view, the atom of radium would be in a process of evolution, and we should be forced to abandon the theory of the invariability of atoms, which is at the foundation of modern chemistry. (Curie, 1904, 461-6)

A June 26, 1903, article in the *Times Literary Supplement* gave a sense of the significance of Mme. Curie's discovery: "It seems as if we have reached an entirely unknown region in physics, where all analogies fail us, and the accepted views of the nature of matter offer no assistance."

CHAPTER 7

Yes, It Is Transmutation!

Soddy and Ramsay Destroy a 100-year-old Assumption and Prove Natural Elemental Transmutation (1903)

In July 1903, four months after Pierre Curie and Albert Laborde published their seminal paper on the powerful heat released by radium, John Dalton's 1808 model of an indivisible atom was laid to rest. Two scientists in London performed an experiment that proved Dalton wrong, with their observation of a natural transmutation of an element.

Ramsay and Soddy Make History

The first instance on record, according to historian Ida Freund, of this transmutation took place when Frederick Soddy and William Ramsay observed that helium was produced by the radioactivity of radium. (Freund, 1904, 619)

Soddy, the chemistry brainpower of the Rutherford-Soddy duo, had recently left McGill University and gone to London to work with Ramsay. Rutherford and Soddy certainly knew that they had transmuted thorium into one of the argon-type gases, but they didn't make a direct claim of transmutation in their journal articles, nor did they specifically identify the produced element.

Their conservative approach may have been wise, considering the possible resistance they faced in light of the established model of Dalton's atom. Although science encourages its practitioners to make big discoveries, it normally does not relinquish long-accepted philosophies without a fight and without conclusive evidence. This battle, for the new philosophy of a divisible atom, was relatively bloodless, however, as science historian Lawrence Badash explained.

"In July of 1903," Badash wrote, "Ramsay and Soddy produced the most impressive single piece of evidence in confirmation of the theory that radioactive atoms decayed into other elements. Using liquefied air, they condensed the gaseous emanation from radium, and by spectral examination of the residue, they discovered that it contained helium. There could be little doubt that the helium was a breakdown product from the radium — a clear case of one element giving rise to another." (Badash, 1966, 92)

The Simple Spectrometer

Ramsay and Soddy used a common and important diagnostic tool, called the optical spectrometer, to identify the helium. It was a remarkably informative tool that allowed identification of elements simply, precisely, quickly and easily.

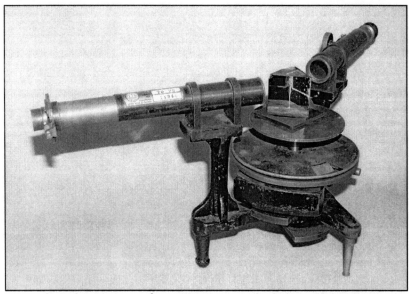

Optical spectrometer from 1896. Image credit: Indiana University

Although the spectrometer is not as precise as modern-day electronic spectrometers, the historical spectrometer was virtually foolproof, and its results were difficult to dispute.

The apparatus consisted of two tubes with a prism between them. An illuminated source, such as a burning flame of gas, or an electric-arc (like a fluorescent light), was placed at the end of one tube. The light emitted from the source traveled down the tube and passed through the prism. The user of the device looked through the end of the other tube, toward the prism.

Prisms placed in direct sunlight receive a beam of sunlight on one side and, on another side, emit and spread out that beam, revealing that sunlight comprises a full rainbow of colors. This familiar rainbow begins with ultraviolet on the left and ends with infrared on the right. It's known as a full spectrum because sunlight is composed of a set of continuous, virtually infinite gradations of every color in the rainbow. Diffraction grating, tiny parallel etched lines on a flat surface, like the underside of a CD-ROM or DVD-ROM, produces the same effect. Nowadays, simple spectrometers suitable for classroom use are available for $15 in scientific catalogs.

Full spectrum of visible light within the electromagnetic spectrum. Image courtesy of Cornell University (Colors will not be visible in all editions of this book.)

When a gas is heated by combustion or made to glow by an electric current passing through it, the gas emits a pattern of light that is distinct to that element — essentially, a visible fingerprint of the element. Shown below are spectral images of four gases. The longer line for each gas, glowing inside a tube, stimulated electrically by a 5,000-volt transformer, is the color that is visible to the naked eye. The colored lines in the inset show each gas as viewed through a spectroscope. (Colors will not be visible in all editions of this book.)

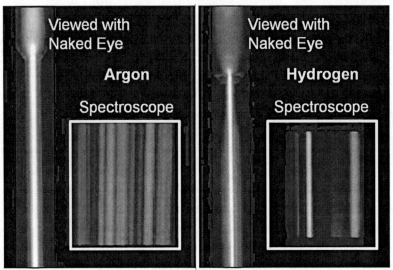

Argon spectrum (L), Hydrogen spectrum (R)
Images courtesy of Georgia State University

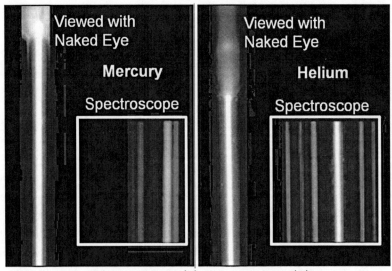

Mercury spectrum (L), Helium spectrum (R)
Images courtesy of Georgia State University

Ramsay Talks Transmutation

Ramsay and Soddy's four-page July 1903 paper that introduced their transmutation claim was short and simple. It explained what they did, how they did it, and how they made their measurements. There was no theory, and it didn't even mention transmutation.

Ramsay and Soddy had definitively proved elemental transmutation and were the first to observe a natural transmutation from an unstable element. This was still not a transmutation in which the hand of man demonstrated the as-yet mythical power to change elements. Ramsay and Soddy's role in the transmutation was passive: They simply did the measurements and analysis of a naturally radioactive source material.

On Nov. 26, 1903, Ramsay gave a lecture at the London Institution and, according to the *New York Times*, spoke a bit more freely than he and Soddy had done in their paper. Not only did he make clear that they had observed an elemental transmutation, but he — in what certainly raised the hackles of his more-conservative colleagues — also mentioned that the transmutation of metals was not, after all, so absurd a theory.

> **Another Discovery as to Radium.**
> LONDON, Nov. 26.—Sir William Ramsay, the celebrated chemist, in the course of a lecture delivered here to-night, described a number of experiments made by him which resulted in the discovery that the gaseous emanation from radium is really helium. From this discovery, Sir William said, it might be concluded that the transmutation of metals was not, after all, so absurd a theory.

Brief news story from the New York Times, *Nov. 27, 1903*

The following day, the *New York Times* ran another, much longer story and was on the defensive. Alchemy, despised by the scientific elite, had reared its head again. The *Times* wrote that the Ramsay-Soddy radium-to-helium experiment "does not prove that the dream of alchemy has become a reality."

A week later, on Dec. 8, the newspaper was still in damage-control mode. "Those who became hysterical over the telegraph reports," the *Times* wrote, "of the recent lecture of Sir William Ramsay, in which he was reported to have announced observations pointing to the transmutation of metals in the change of the emanations from radium into helium, will be interested to know that he has discovered nothing and said nothing which could warrant the belief that he thought he had learned the secret sought by the alchemists of changing base metal into gold." The *Times* reiterated that the claim was only the production of helium. Ramsay clearly knew what he wanted to accomplish next. Meanwhile, the Rutherford and Soddy team, and then later as individuals, was developing one of the most significant concepts of atomic science, which they called the "Law of Radioactive Change."

> **RADIUM AND HELIUM.**
> Those who became hysterical over the telegraphed reports of the recent lecture of Sir WILLIAM RAMSAY, in which he was reported to have announced observations pointing to the transmutation of metals in the change of the emanations from radium into helium, will be interested to know that he has discovered nothing and said nothing which could warrant the belief that he thought he had learned the secret sought by the alchemists of changing base metal into gold.

Top portion of news story from the New York Times, *Dec. 8, 1903*

CHAPTER 8

Giving Credit Where It's Not Due

Soddy Describes the Disintegration Theory; Rutherford Gets the Credit and the Nobel Prize (1903-1904)

Historians have given the team of physicist Ernest Rutherford and chemist Frederick Soddy joint credit for the idea that radioactivity changes elements from one to another, transmuting them by way of radioactive decay. (Trenn, 1977, 5-6)

Science historian Lawrence Badash wrote that Rutherford and Soddy "announced their theory of the transmutation of elements, a discovery that genuinely overturned men's ideas about the nature of matter." (Badash, 1996) They made no such announcement of a transmutation of elements; this is a myth.

Together, Rutherford and Soddy proposed that radioactivity caused elements to disintegrate. Alone, Soddy argued that the disintegration of radium to helium was not merely a separation of its constituent parts but a transmutation from one element to another.

Among their nine papers together, Rutherford and Soddy never jointly published anything about a Disintegration Theory or elemental transmutations.

The specter of alchemy haunted the scientific community at the time. Explicitly stating that the experiments were creating *new elements* and that *transmutations* were occurring was a tremendous political risk; the courage for doing so and the credit belong not to the Rutherford and

Soddy team but to Soddy alone.

When Rutherford and Soddy were working together in 1902-03, as described in Chapter 5, they figured out that something was coming out of their thorium that they detected by its level of radioactivity. They called it an "emanation" and a "non-thorium type of matter." Rutherford and Soddy did not identify the new element or claim to have transmuted elements.

In a 1902 paper, under the heading, "General Theoretical Considerations," the two scientists suggested this preliminary idea:

> Radioactivity is at once an atomic phenomenon and the accompaniment of a chemical change in which new kinds of matter are produced. The two considerations force us to the conclusion that radioactivity is a manifestation of subatomic chemical change. (Rutherford and Soddy, 1902, Compounds II, 859)

"New kinds of matter produced" is not the same as "changing one element into another." Although transmutation is one valid interpretation of what they wrote, it is but an interpretation, and it is not what they said.

Experimentally and in support of their idea, Rutherford and Soddy said only that thorium changed to thorium-X, uranium changed into uranium-X and radium changed into radium emanation.

In May 1903, Rutherford and Soddy published their Law of Radioactive Change in their paper "Radioactive Change." It defined the rate of change when elements decay into an unidentified emanation as a result of radioactivity. They had accumulated, they wrote, sufficient experimental evidence "to enable a general theory of the nature of the process to be established with a considerable degree of certainty and definiteness." (Rutherford and Soddy, 1903, 577-8) They came right up to the edge of saying that their theory suggested the transmutation of elements:

> A body that is radioactive must *ipso facto* be changing, and hence it is not possible that any of the new types of

radioactive matter, that is, uranium-X, thorium-X, the two emanations, be identical with any of the known elements. (Rutherford and Soddy, 1903, 578-9

With great caution, they suggested that one of these changes was producing the element helium:

> We have already suggested on these and other grounds that possibly helium may be such an ultimate product, although, of course, the suggestion is at present a purely speculative one. (Rutherford and Soddy, 1903, 579)

In early 1903, Soddy left Rutherford's laboratory and went to work with chemist William Ramsay.

In July 1903, Soddy and Ramsay definitively established, for the first time, proof of a natural (as opposed to man-made) nuclear transmutation: Radium, they showed, was spontaneously changing into helium. (They did not use the term "nuclear" at the time.) Now the experimental evidence for elemental transmutation existed.

In September 1903, Soddy presented a conference paper and began using the term Disintegration Theory. Not only did Soddy's definition of the term include the concepts that he and Rutherford developed in their Law of Radioactive Change, but also Soddy went further: He said that his Disintegration Theory involved elemental transmutation.

These facts were misattributed as early as 1908, when professor K.B. Hasselberg, president of the Royal Academy of Sciences, awarded Rutherford (but not Soddy) the Nobel Prize in Chemistry:

> As will be seen, Rutherford's discoveries led to the highly surprising conclusion that a chemical element, in conflict with every theory hitherto advanced, is capable of being transformed into other elements, and thus in a certain way it may be said that the progress of investigation is bringing us back once more to the transmutation theory propounded and upheld by the alchemists of old.
>
> As an explanation of these remarkable phenomena,

> Rutherford, in conjunction with Dr. F. Soddy, one of his numerous co-investigators, brought forward in 1902 the so-called disintegration theory, which is closely allied in several aspects to the opinions earlier enunciated by J.J. Thomson and other physicists with regard to the nature of matter. (Hasselberg, 1908)

Rutherford may have known that elements were being transmuted into other elements through radioactivity; however, as shown in Chapter 5, Soddy told his biographer that Rutherford had denied to Soddy that elements had transmuted in their experiments. Rutherford was keenly aware of the political risk of making a claim of transmutation. He took pains to avoid associating his work with changing elements — that is, alchemy.

The purposes of this chapter, therefore, are to unscramble a tangled scientific history, trace the history of the Law of Radioactive Change and the Disintegration Theory as well as properly attribute credit. Some additional information about radiation will be helpful.

Types of Radiation

These pioneers of atomic science were able to gain a remarkable level of knowledge about the characteristics of radioactivity, considering that no aspects of the phenomenon were directly observable with any of the human senses.

By this time, the international group of radioactivity researchers knew that radioactivity was accompanied by three types of radiation. The first two were labeled with the Greek characters "alpha" and "beta". Researchers later figured out that alpha particles are helium nuclei and beta particles are energetic electrons. The third type of radiation was the gamma ray. Gamma rays are highly penetrating forms of electromagnetic radiation emitted from nuclear reactions. On Earth, they are emitted by devices or radioactive materials and a few rare terrestrial events. Gamma rays are a class of photons (a larger group of massless entities) that, according to quantum mechanics, behave both as waves and as particles. A range of various-energy gamma rays can be depicted in a gamma spectrum.

Researchers also understood the concept of radioactive decay, which describes how radioactive elements spontaneously disintegrate and emit alpha particles or beta particles. As radioactive elements decay, the alphas and betas fly from them into the surrounding space. In their place, they leave smaller and slightly different elements or isotopes. Gamma radiation is different in nature and is not a mechanism for radioactive decay, as alpha and beta particles are. For more information on the types of radiation and their characteristics, see Appendix A, Emissions From Spontaneous Radioactivity.

> Note: From time to time I will present optional short sections with technical information for readers who are curious about the deeper scientific details. These sections will appear with shaded background like this.

Law of Radioactive Change

By May 1903, Rutherford and Soddy had published nine papers together and examined the nature of the radioactivity from three elements: thorium, uranium, and radium. They noticed a factor common to all three: As each element decayed — when the strength of the radioactivity of an element decreased over time — they simultaneously detected the presence of something new, unidentified but measurable by its increasing radioactive strength. They also noticed that the rates of the disintegration, or decay, were universal. They therefore established the Law of Radioactive Change, which is expressed in the graph below.

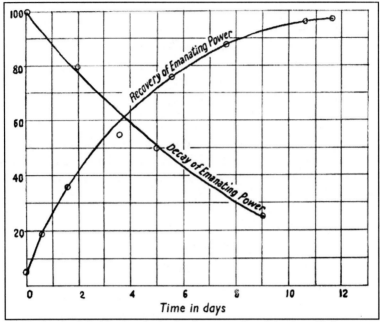

Rutherford-Soddy graph: As the first samples of thorium-X decay, new samples of thorium-X are produced.
(Rutherford and Soddy, 1902, Compounds I; Compounds II)

Some information would be helpful to understand the graph from Rutherford and Soddy's perspective a century ago. When they published this graph showing plots of the increasing and decreasing radioactivity, they were working with thorium. It was far too early in the history of nuclear science for them to know what was going on at an elemental level, let alone at an isotopic level. The term isotopes hadn't even been invented.

Here is an explanation of the curves. The parent, thorium, begins to decay to the child, thorium-X. The radioactivity in the first thorium-X sample has a half-life of about four days and quickly loses most of its radioactivity. At the same time, in the parent thorium sample, thorium-X continues to be produced. The decay curve shows the reduction of activity in the produced thorium-X, and the recovery curve shows the simultaneous increase of new thorium-X activity in the parent, the thorium.

Nucleosynthetic Reaction Network

Rutherford and Soddy recognized that the changes were not occurring in simple, two-step processes. In fact, the reaction path between their starting material, thorium, and their detected material, thorium-X (later known as radon), could have involved 11 unique steps. These steps are called nucleosynthetic reactions, and the transmutations and isotopic changes occur through alpha and beta decay processes.

Radioactive material can decay through an alpha decay, a beta decay, or both, as well as a few other types of decays. In these nucleosynthetic reactions, the starting material is called the "parent," and as it decays, a new material is born; this is called the "daughter" element. (More precisely, elements depicted in nucleosynthetic reactions are usually identified by their isotopes.) When that "daughter" element begins to decay, it is considered a "parent" to its own "daughter" element, and so on. The reaction path continues until it comes to a natural stop by decaying into a stable element. To simplify things, here is a snapshot of a two-step process in the middle of a much longer reaction network.

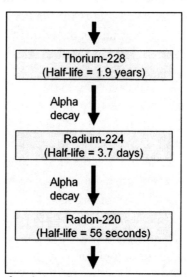

Portion of nucleosynthetic reaction network depicting change from thorium-228 to radon-220.

New Law, in Their Own Words

The original text in which Rutherford and Soddy explained their law is difficult to follow because it represents the first scientific descriptions of newly observed phenomena and the terminology differs from what we use today. However, it is important because it shows how scientists formulated their views regarding newly discovered phenomena. Here is what Rutherford and Soddy wrote:

> When several changes occur together, these are not simultaneous but successive. Thus, thorium produces thorium-X, the thorium-X produces the thorium emanation, and the latter produces the excited activity. Now, the radioactivity of each of these substances can be shown to be connected, not with the change in which it was itself produced but with the change in which it in turn produces the next new type. Thus, after thorium-X has been separated from the thorium producing it, the radiations of the thorium-X are proportional to the amount of emanation that it produces, and both the radioactivity and the emanating power of thorium-X decay, according to the same law and at the same rate. In the next stage, the emanation produces the excited activity.
>
> The activity of the emanation falls to half-value in one minute, and the amount of excited activity produced by it on the negative electrode in an electric field falls off in like ratio. These results are fully borne out in the case of radium. The activity of the radium emanation decays to half-value in four days; so does its power of producing the excited activity. Hence, it is not possible to regard radioactivity as a consequence of changes that have taken place. The rays emitted must be an accompaniment to the change of the radiating system into the one next produced. (Rutherford and Soddy, 1903, 578, Radioactive Change)

Remarkably, considering the technical limitations Rutherford and Soddy faced, they understood that radioactive material, by nature's own processes, eventually becomes nonradioactive and harmless.

> On the other hand, since the ultimate products of the chain cannot be radioactive, there must always exist at least one stage in the process beyond the range of the methods of experiment. For this reason, the ultimate products that result from the changes remain unknown, the quantities involved being unrecognizable, except by the methods of radioactivity. (Rutherford and Soddy, 1903, 579, Radioactive Change)

Unscrambling History

Rutherford was reluctant to suggest that they had created new elements; rather, he preferred the more ambiguous term "emanation." Later, Soddy, while working with Ramsay, looked in greater detail at the chemical changes and proved that these chemical changes were in fact elemental transmutations. (Radioactivity also causes isotopic changes, but they hadn't gotten that far yet.)

Rutherford and Soddy's work earned Rutherford the Nobel Prize in Chemistry in 1908 "for his investigations into the disintegration of the elements, and the chemistry of radioactive substances." Nobelprize.org, the official Web site of the Nobel Prize, says that Frederick Soddy "collaborated with Rutherford in creating the disintegration theory of radioactivity, which regards radioactive phenomena as atomic — not molecular — processes."

But there is no such disintegration theory of radioactivity published by Rutherford and Soddy. They didn't mention anything called a disintegration theory in their papers. Among their nine joint papers, only the last one, "Radioactive Change," published in May 1903, even mentions the word disintegration. Their paper "Radioactive Change" is the seminal paper in which they published their Law of Radioactive Change and the basis on which Rutherford won the Nobel Prize.

The first instance I found of the term "disintegration theory" is in a June 26, 1903, newspaper article titled "The Disintegration Theory of

Radioactivity," in the *Times Literary Supplement.* (*Times Literary Supplement*, 1903, 201)

The article says that "the disintegration theory was put forward a year ago by Professor Rutherford and Mr. Soddy, working in Montreal, to explain the radioactivity of thorium, and it has since been extended by them to include the behavior of radium and thorium as well." The writer was brilliantly and technically accurate in describing the research but appears to have confused the terminology.

Rutherford and Soddy had offered the Law of Radioactive Change, not the Disintegration Theory. The writer described "the disintegration theory of radioactivity" in his own words: "It postulates that the atoms of radioactive matter are so constituted that sooner or later in their history they become unstable systems and fly apart into smaller fragments." A few months later, Soddy described his Disintegration Theory, perhaps borrowing the term from the reporter or common use at the time.

The Disintegration Theory

In the fall of 1903, Soddy began lecturing and writing about his insights from his work with Rutherford and Ramsay. Soddy began to connect the dots and see the relationships among radioactivity, the process of radioactive decay and elemental transmutation.

On Sept. 18, 1903, a reporter for *The Electrician* wrote about the Sept. 10, 1903, meeting of the British Association for the Advancement of Science:

> A paper on the emanations of radioactive bodies was read by professor Rutherford. He described the discovery of emanations, sketched their chief properties, and in fact summarized our present knowledge of emanation phenomena. He placed before the meeting, very clearly, the theory of radioactivity which he favors. This theory, stated briefly, affirms that some of the atoms of radium, becoming unstable, fling off a part of themselves positively charged,

leaving atoms of a new substance, the emanation. The atoms of the emanation behaved similarly; and thus, perhaps by several stages, the stable element called helium is finally reached. (*Electrician*, 1903, 880)

This description is consistent with Rutherford and Soddy's papers. It does not use the term "disintegration theory," and it says that helium is the byproduct of disintegration, not transmutation.

Rutherford's paper was not accepted without a fight. After his presentation, Sir Oliver Lodge read a prepared statement from Lord Kelvin (William Thomson), who disputed Rutherford's disintegration hypothesis. Kelvin, according to the reporter, attributed the phenomena to the absorption of stray ether waves.

Henry E. Armstrong, a professor of chemistry at Central Technical College, in London, was next to argue against Rutherford's disintegration theory of radioactivity. Speaking on behalf of chemists, Armstrong said they were astonished at the physicists' feats of imagination. Soddy said that Armstrong's astonishment was exaggerated.

Then Soddy used and defined the term "disintegration theory," perhaps for the first time. The reporter described Soddy's statement:

> Mr. Soddy thought the novelty of the disintegration theory of radioactivity had been exaggerated at the expense of its probability. It contains nothing contrary to the fundamental laws of chemistry. The theory supposes that a gradual transmutation of the radium occurs because individual atoms of the mass make a sudden change. ... In all radioactive phenomena, two things happened: the emission of rays and the production of the new substance. The amount of new substance produced was proportional strictly to the number of rays emitted. (*Electrician*, 1903, 880)

Soddy's biographer, Muriel Howorth, described how vociferously Soddy defended his understanding of the phenomena:

> Rutherford himself seems to have been rather apprehensive — and with some reason — for Armstrong sided to a certain extent with [Lord Kelvin]. In fact, [Armstrong] rashly tried to ridicule the whole theory, saying that chemists certainly had no evidence of atomic disintegration on the Earth. Soddy, in his indignant reply, was so vehement that the chairmen had to intervene. (Howorth, 1958, 91)

But Soddy had seen the evidence firsthand. Soddy had gone further than he and Rutherford had. Together, they had identified the reaction product as an "emanation," a "non-thorium type of matter" and a new unspecified substance. With Ramsay, Soddy had proven that the "emanation" was the newly transmuted element helium.

A month later, on Oct. 13, 1903, Soddy enthusiastically began a six-week series of 12 lectures at University College on radioactivity. His lectures included material from a book on radioactivity that he had nearly finished. (*Electrician*, 1903, 992) Between Oct. 23, 1903, and Feb. 19, 1904, Soddy published 17 articles in *The Electrician*, based on his forthcoming book. The editors of *The Electrician* introduced Soddy's series as "what may practically be regarded as a textbook on radium and radioactivity." (*Electrician*, 1903, 1) It would have been the first such textbook, but Rutherford interfered. Rutherford's biographer, Arthur Eve, explained what happened.

"When Soddy reached England," Eve wrote, "he was full of enthusiasm and gave a series of public lectures which were fully recorded in *The Electrician*, and it was feared that these might damage the sale of Rutherford's book. Soddy also undertook to write a book on "Radioactivity" and, although he proposed to emphasize the chemical side of the subject, there is much material which he had obtained from his college at McGill at lectures and otherwise. There is no doubt that Rutherford felt strongly on this question, and he wrote a firm letter expressing his point of view. It was finally agreed that Soddy's book should not appear until one or two months after publication of Rutherford's." (Eve, 1939, 99)

The first edition of Rutherford's book *Radio-activity* published in

February 1904. Soddy's book, *Radio-activity: An Elementary Treatise From the Standpoint of the Disintegration Theory*, published in May 1904. As indicated by the title, Soddy's book was written specifically from the perspective of the disintegration theory, which, as he said, was intrinsically related to elemental transmutation.

Soddy's insight into atomic disintegration led him to develop the concept of isotopes and apply the term, for which he was awarded the 1921 Nobel Prize in Chemistry. Additionally, he was the first person to propose that stable elements, not just the radioactive ones, might have different isotopes. (National Academy of Sciences, 1966)

To recap, Rutherford and Soddy published what they called the Law of Radioactive Change. They never published anything called a Disintegration Theory. Soddy alone published a book on the Disintegration Theory.

To many Rutherford supporters, the implication that Rutherford may have had any responsibility for, or even permitted, scientific credit to be mis-apportioned is tantamount to blasphemy.

That very well may have happened; it was Soddy who specifically intended to transmute elements, who first told to Rutherford that they had done so, and who first had the courage to say publicly that they had done so.

This is not, however, the only situation in which Rutherford appears to have been given — or taken — unwarranted credit. Chapter 23 describes another matter of mis-apportioned credit; in that instance, the published scientific record leaves no ambiguities.

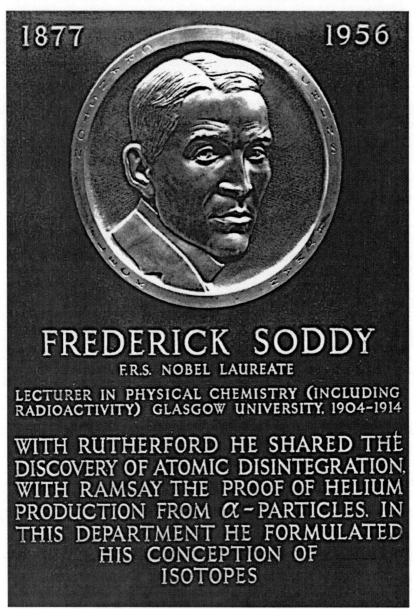

Soddy plaque at Glasgow University

CHAPTER 9

From LSD to Alchemy

One Man Found It: the Lost Trail of Early Transmutation Research

Up to this point, we have discussed the formative period of atomic science, which began with Roentgen's discovery of X-rays in 1895, through Ramsay and Soddy's helium transmutation discovery in 1903. That research is part of the well-known history of science. But now we enter unfamiliar territory, traveling to a hidden period of transmutation research. I did not find this hidden period on my own. A young man who traveled from Florida to California in the 1960s helped me find the trail.

As a teenager, Robert A. Nelson got into trouble often. His parents punished him by keeping him at home. To occupy his active mind, he read through every page of the family's copy of *Encyclopedia Americana*. When he was 15, his parents disowned him, and he spent a year in juvenile hall. In 1966, he escaped and hitchhiked across the country to Berkeley, California, at the peak of the Summer of Love hippie revolution.

"One day as I was walking up Telegraph Avenue," Nelson told me, "the thought occurred to me that I could make a lot of money if I learned how to make LSD. So I walked up to the biochemistry library and started looking up lysergic acid in *Chemical Abstracts*. The rest is a long story that includes 13 years of underground chemistry, during which time I had recall of a former life as the alchemist Joseph Duchesne."

Nelson went from learning how to manufacture LSD to studying alchemy. His journey also brought him to the world of alternative energy, science and technology. After reading *Suppressed and Incredible Inventions*, written by John Freeman and published by Al Fry, Nelson was relentlessly persistent and went on a hunt to collect as many of Fry's publications as he could find. Most of it existed only as faded photocopies. He eventually met Fry in Perris, California, and Fry told Nelson that he was getting tired of publishing, that he wanted to pass the work on to someone else. Nelson was thrilled and took up the torch in 1982, publishing by making photocopies.

In addition to his interest in making LSD, Nelson was fascinated by occult science, including alchemy. He compiled and assembled an encyclopedic, academically referenced collection of information in his book *Adept Alchemy*. In 1998, he self-published the book by making photocopies and, sometime before May 10, 2006, according to the Internet Archive, published an online version of it.

Nelson meticulously documented important nuclear transmutation work performed during the early 20th century as well as a host of other fascinating topics. It is a wondrous compilation of science, quasi-science, mysticism and magic. He created an interlinked, contextual reference guide to this lost collection of research.

Nelson appears to be the only person who has found and documented this lost period of science history — at least in the English-language — since the 1930s. He has done so with remarkable insight.

Nelson read each paper, discussed the technical details of the experiments, and wrote a summary of the results. He kept his personal speculation and judgment to a minimum. As far as subject matter, he was inclusive rather than exclusive, and as a result, some of his references likely would not pass scientific muster. However, a lot of the research Nelson identified was published in top-tier journals, including *Physical Review, Nature, Science* and *Journal of the American Chemical Society*. Some of this research was performed by Nobel Prize winners.

How someone like Nelson, without a scientific background, was guided and able to follow the interconnected threads of this research is a mystery. Perhaps he has a photographic memory and he was able to keep a mental map as he was leafing through the archives from nearly a century ago.

74 • FROM LSD TO ALCHEMY

Preceding page: "Alchemia," designed and commissioned by Nelson and drawn by Hal Robins. The drawing summarizes Nelson's understanding of alchemy from this early period. I stumbled on Nelson and his work during my search to find out more about these studies and their possible relationship to modern nuclear transmutation research.

My investigation into the early 20th century origins of nuclear transmutation research would not have been possible without Nelson's work. I had known of only a few scientific papers about the work between 1912 and 1927. Nelson's bibliography and collection was my guide to the rest of them. He showed me the broad landscape of this subject, and I went to the original journal articles and newspapers to obtain my own, first-hand account and understanding of this history.

As I dug deeper into the material, I was amazed that so much of it had been omitted from science textbooks and forgotten in science history. But I understood why: This early transmutation research made no sense according to theory at the time and was poorly reproducible. On the other hand, with the benefit of hindsight and the knowledge of modern low-energy nuclear reaction research, the work from that period made sense to me. The research also shows that chemists had a more significant role, however unappreciated, than history has acknowledged.

Not the Sincerest Form of Flattery

As an editor of this book was checking facts, we came across a 2014 book published by Oxford University Press containing one section that broadly discussed many of these early transmutation experiments.

The book is *The Lost Elements: The Periodic Table's Shadow Side,* by Marco Fontani, Mariagrazia Costa and Mary Virginia Orna. Fontani and Costa are chemists and work at the University of Florence, Italy. Orna is a professor of chemistry at the College of New Rochelle, New York.

The three authors appear to have done an apparently scholarly review of the topic, particularly because they are not specialists in low-energy nuclear transmutation research. Their treatment of the topic is largely superficial. However, this is understandable in light of the fact that the research to which I have devoted an entire book is discussed in only 33 pages of their book, in the section called "Modern Alchemy: The Dream to Transmute the Elements Has Always Been With Us."

A major limitation of their review of the subject matter is that they spent very little time examining the original scientific papers. As a result, they can't provide readers with the detailed facts that support a view that, for at least some of the research, the results were real. The three authors assist their readers in concluding that all the research was invalid by writing that other scientists who attempted the experiments failed and that such failures allegedly negated the reported successes. This, however, is an erroneous and invalid approach toward scientific skepticism.

As I was reading through the Fontani-Costa-Orna book, I found a few anomalies. Most notably, the authors mention J.J. Thomson's observation of the as-yet-undiscovered hydrogen isotope that is now called tritium. (Chapter 13 in this book.) I was puzzled that the authors failed to recognize this as clear evidence of a nuclear transmutation.

But then I noticed serious problems in the Fontani-Costa-Orna book. Their transmutation section duplicates not only the broad scope of Nelson's book but also whole paragraphs of Nelson's original book without attribution to Nelson.

The authors were aware of Nelson's book, but in the more than 100 references in their "Modern Alchemy" section, they cite his 1998 book only once. They list his book in their bibliography, and they say nothing more about Nelson. It gets worse.

In many of the paragraphs I compared, where Nelson had given facts in *Adept Alchemy* that supported the validity of these transmutations, in *Lost Elements*, Fontani, Costa and Orna omitted or changed specific words to steer readers away from concluding that some of the 1912-1927 transmutation experiments were valid.

I do not know whether all three authors were involved in the plagiarism. I have not examined other sections of the book for similar problems. Among the more than 100 named individuals in the acknowledgments, Fontani, Costa and Orna did not mention Robert

Nelson, nor did the authors identify any assistants who did the research and writing on their behalf. I have published a detailed comparison of the two books on the *New Energy Times* Web site. The comparison is listed under the *Investigations* heading.

CHAPTER 10

Ramsay Breaks the Ice

Chemist's Attempt at Man-Made Transmutation Offers Lessons on the Scientific Method (1907)

Before we dive into the murky waters of the early 20th century alchemy attempts, let's do a quick review. By the time the science of radioactivity arrived in the late 1800s, the scientific world "had come to the conclusion that it knew pretty nearly all there was to be known," according to the *New York Times* in 1911. (*New York Times*, 1911)

In the early 1900s, Rutherford, Soddy and Ramsay had revealed that spontaneous nuclear transmutations — by natural processes — occurred on Earth. This revelation was generally accepted without uproar, thanks to three factors: 1) the bulk of the work was experimental rather than theoretical; 2) each successive experiment built on the previous ones; and 3) Rutherford introduced his and Soddy's disintegration claim (distinct from Soddy's Disintegration Theory) shrewdly.

In 1902 and 1903, Rutherford had carefully guided his and Soddy's research papers through the rocky waters of review and publication. He had an outstanding reputation among his peers and the international leaders in physics academia. He also had clout, and he was politically sophisticated. Ramsay and Soddy's 1903 independent confirmation of observing helium from radium added strength to Rutherford and Soddy's disintegration claim. Other scientists were able to easily reproduce their experiments. Acceptance of the phenomena of radioactivity and natural transmutations had arrived.

In 1905, Albert Einstein published his theory of special relativity: the relationship between mass and energy, $E=mc^2$, which states that energy is proportional to mass times the speed of light squared. During this time, several people were developing models of the structure of the atom. By 1907, the study of radioactivity was a hot topic among chemists and physicists, but it was exclusively owned by neither discipline.

We now approach unstable ground in our journey: the first claims of man-made elemental transmutation. Scientists were not content simply to watch a radioactive element disintegrate (decay) spontaneously into another element; they wanted to deliberately transmute one element to another.

The hypothesis that mere mortals could wield the power to effect elemental transmutations at will was heretical; it bore the taint of alchemy. Few reputable scientists were willing to risk their careers by having, or even appearing to have, any association with alchemy. They knew that the alpha particles emitted from radioactive materials had immense energy and that, if anything had the power to make such changes, alpha particles did. The idea, however, was still heretical.

The First Attempt

In 1907, one scientist had the courage as well as the political stature to attempt a man-made transmutation: Sir William Ramsay (1852-1916.) In 1904, he had been awarded the Nobel Prize in chemistry, and he became president of the Society of Chemical Industry. According to one of his biographers, William Tilden, Ramsay was honored 90 times for scientific achievements during his lifetime. (Tilden, 1918, 307-308)

Some of the starting materials he used were water, copper sulfate and copper nitrate. Some of the newly created elements he claimed were lithium, sodium, helium, neon and argon. After an extensive review of the original scientific papers and half a dozen historical accounts of this claim, I have been unable to determine whether Ramsay succeeded in his transmutation attempts. Nevertheless, other aspects of his work are important.

In this set of experiments, Ramsay was the first scientist to try to use the energy of alpha particles to initiate man-made elemental transmutations in stable elements. He was also the first prominent scientist to claim that he had performed a man-made transmutation from stable elements. His courage to explore a new boundary of science inspired other scientists to develop curiosity in similar experiments in the next few years. These later experiments — and their successful results — foreshadowed other chemistry-based transmutation research that would take place 70 years later.

Sir William Ramsay, circa 1912

These experiments from the early 1900s appeared, among other places, on the front page of the *New York Times* and were reported internationally. Only a few very old history books have mentioned these early transmutation experiments. They do not seem to appear in any modern accounts of science history.

Particularly because the result was ambiguous, the 1907 Ramsay transmutation claim provides an excellent example with which to examine the scientific process and the process by which scientific claims are adjudicated by the scientific community.

Failure to Confirm

Nowadays, nuclear transmutations using high-energy physics are accepted without question because the results are easy to reproduce and because they fit prevailing theory. However, until the experimental and theoretical advances that emerged between 1998 and 2005, as discussed in the book *Hacking the Atom: Explorations in Nuclear Research, Vol. 1,* any scientist or scholar who supported the idea of transmuting elements without high-energy physics risked appearing unscientific, if not a fool.

Mark Morrisson is one of the most recent scholars to look at the relationship between alchemy and the early transmutation research. Morrisson is a professor and the former head of the English Department at Pennsylvania State University. He wrote the book *Modern Alchemy: Occultism and the Emergence of Atomic Theory* in 2007. In his book, Morrisson discussed some of the topics in this book. He too, learned about some of the lost research from Robert Nelson's Web site. Morrisson discussed a few of the man-made transmutation attempts and superficially dismissed each of them as a mistake or error.

Specifically, he reported that other scientists who had attempted to repeat these experiments failed to obtain positive results. Morrisson implied that these failures negated the positive results.

He provided readers with statements from critics that reinforced his depiction of the reports as mistakes, frauds, or hoaxes. He also failed to examine the work of Wendt and Irion, Nagaoka, Paneth and Peters

(Chapters 18, 21 and 24 in this book).

On the Matter of Scientific Repeatability

A scientist must be able to successfully repeat his or her experiment; this is a fundamental part of the scientific method. Once a scientist satisfactorily completes the steps of the scientific method, he or she must share sufficient details of the experiment so that it may be repeated by another scientist or, preferably, by many scientists. As a condition of acceptance of a new claim, the scientific community requires that an experiment be generally repeatable.

In his book *Changing Order: Replication and Induction in Scientific Practice*, Harry M. Collins writes about the occasionally difficult task of knowledge transfer from one scientist to another who is attempting to repeat an experiment.

There are, however, innumerable procedural details and subtleties that can be required to successfully repeat an experiment. A failure to communicate such details can lead directly to a failure to confirm an experiment. Replication, however, can be limited by unscientific factors.

Scientists are subject to the same human irrationalities and subjectivity that the rest of us are. Factors that come into play can include ego, prestige, and fame. With any potentially important scientific claim, there is always competition for patents, funding, resources and Nobel prizes. Perhaps the most significant adversarial factor is the battle between prevailing and emerging ideologies.

There is a practical side to established scientific principles, though. Science gains its strength from a foundation of knowledge built gradually and carefully, one step at a time. Time-tested scientific principles should not be displaced whenever one person proposes a new idea, although valid experiments, according to the scientific method, always trump theory.

Given the technical challenges of replicating radically new and potentially important experiments, and given the myriad potential human factors, repeatability challenges can be appreciated better with some consideration of human nature. If a scientist stands to lose funding and prestige if he or she successfully repeats another scientist's experiment, then he or she has an inherent conflict of interest. If a scientist, before making such an attempt, has gone on record opposing the new idea, the scientist likely won't have the necessary objectivity. The same conditions apply if the person is a competitor of the originator's. In major unresolved science explorations, rivalries have usually been long-established.

Scientists are expected to do many things to help their peers assess the validity of their claims. Included among these things are showing data, explaining procedures and stating assumptions. Underlying all of this is an implicit assumption of trust. It is virtually impossible to oversee every step and every measurement performed by a scientist in the lab. Trust, therefore, is a foundation for the practice of science. This applies equally to a claimant and a scientist attempting a replication.

There is another factor that can adversely affect the repeatability of the experiment. Scientists attempting to repeat another's experiment often make changes to the experimental protocol, thinking that they are improving it, when, in fact, they have made a crucial change that causes the experiment to fail. This can easily happen in new lines of research.

For these reasons, a failure to successfully repeat an experiment does not provide a valid basis to discredit a novel claim.

The most effective approach to critically evaluate a scientific claim, as author Charles Beaudette told me, is to look for a specific error of protocol, a mistake in the data analysis, or an unstated assumption by the claimant.

Technical Overview of Ramsay's Claims

Ramsay began working on his transmutation experiments between 1905 and 1906, though he did not initially have transmutation as his objective. His initial interest was the effect of radium on water. For most of the work in this series of experiments, he worked with Alexander Thomas Cameron, a postgraduate researcher who studied under Ramsay at University College, London. They published four papers together on this topic, but for simplicity, I will refer to Ramsay as the lead scientist in this chapter.

In 1907, Ramsay made the first set of claims of man-made elemental transmutations. He claimed that he made lithium, sodium, helium, neon and argon. A few months later, two sets of his competitors, Ernest Rutherford and Thomas Royds, and Marie Curie and Ellen Gleditsch, attempted to replicate different aspects of Ramsay's experiments. Both groups reported that they failed to replicate Ramsay's results. Based on these failures, they implied that Ramsay had not achieved what he claimed. Neither group, however, found any fatal flaw with Ramsay and Cameron's work, though Rutherford speculated on one possible source of error.

Main Chemistry Laboratory at University College, London

However, Ramsay's claims were not confirmed by independent replicators. If anyone besides Curie and Rutherford attempted a replication, they did not succeed and publish. Therefore, the scientific record does not definitively confirm or disconfirm Ramsay's claims. Many historians have simply omitted Ramsay's attempted transmutation from their books. Others have implied, citing the Rutherford and Curie groups' publications, that Ramsay's claims were wrong. Regardless of any potential scientific validity, Ramsay's claims marked a significant milestone in the history of elemental transmutation research because of the worldwide public attention and interest they created.

Technical Analysis of Ramsay's Claims

By 1903, several researchers had found that radium could dissociate hydrogen and oxygen when immersed in water. Ramsay performed one of these experiments and, like the researchers who preceded him, found an anomaly. The ratios of evolved hydrogen and oxygen gases did not match the expected ratios. There was always an excess of hydrogen or a loss of oxygen. Ramsay was curious and investigated further. The beginning of his first paper gives more of his perspective:

> The emanation from radium is one of the most potent, if not the most potent, chemical agent which exists in nature. Of all known substances, it is endowed with the greatest content of potential energy: for one cubic centimeter contains, and can evolve, nearly three million times as much heat as an equal volume of a mixture of two volumes of hydrogen with one of oxygen. The spontaneous change which it undergoes, moreover, is accompanied by the emission of an immense number of [particles], expelled with a velocity approaching that of light in magnitude, and which have a remarkable influence on matter. For some years, I have been studying its chemical action, and in this memoir, I will describe its action on pure water. (Ramsay, 1907, 931, Chemical Action)

Ramsay uses the term "action" in these papers, and what he means is the behavior, or the effect, of radium on pure water or solutions containing other dissolved solids. Ramsay thought that, because the radium could separate the hydrogen and oxygen of water, perhaps if he placed a copper sulfate compound in the water, he could segregate the elements and obtain metallic copper. It didn't work; he never got any copper. Instead, in his first set of these experiments, performed in the summer of 1906, he found a trace of lithium in the solution, along with a considerable amount of sodium. "The result was so unexpected," Ramsay wrote, "that it was repeated during the autumn of that year and again in the spring of 1907, always with the same result." (Cameron and Ramsay, 1907, 1594)

Then Ramsay ran another set of experiments and changed a few variables. In some, he found a spectrum of helium and, in others, a spectrum of neon. The helium spectrum was not unexpected, because it is a natural decay product of radium. But the neon was unexpected. In his next experiments, he switched out the copper sulfate for copper nitrate. This time, he saw no helium or neon. Instead, he got a spectrum of argon. None of the control experiments produced any anomalies. (Cameron and Ramsay, 1907, 1600)

In their third paper, Ramsay and Cameron gave voluminous details about their experimental protocol and data. However, they didn't seem to have any definitive conclusion. By the fourth paper, however, they did. "Water and solutions of copper salts were treated with [radon] emanation," Ramsay and Cameron wrote, "and the resulting products [were] examined. It appears that, in [the] presence of water, [radon] emanation disintegrates into neon; [and] with copper solutions, [it disintegrates] into argon. There was some indication that minute traces of copper were transmuted into lithium, and possibly also into sodium." (Cameron and Ramsay, 1908, 992-3, Part IV)

I tried to make a table listing each of the configurations tested and the respective anomalies found in each. I have been unable to clearly follow all of the data presented in Ramsay and Cameron's papers, and thus I cannot present such a table. I have, however, been able to follow the dominant themes of the neon as well as the lithium claims.

Ramsay did not have robust data to support the lithium claim and, accordingly, did not make lithium the central focus of his claim. He also wrote that it was difficult to prove that the argon was not from contamination. In September 1908, Curie reported that she and Gleditsch had attempted a replication of the copper-to-lithium claim. Considering that Ramsay's lithium claim was weak, by his own admission, it's not surprising that Curie failed to obtain positive results. (Curie and Gleditsch, 1908) Some historians have erroneously used the failed Curie replication attempt to imply that, by inference, the Ramsay lithium transmutation was an error. This logic is faulty, as noted earlier.

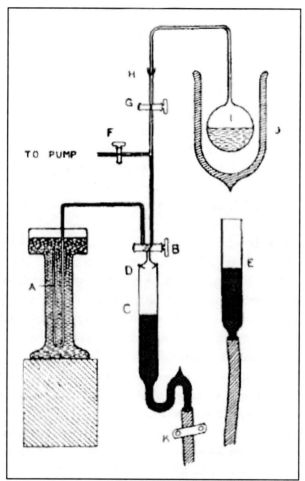

Ramsay-Cameron Experimental Apparatus

In his papers, however, Ramsay based the weight of his transmutation claim not on lithium but on the observation of neon. He was well aware of the possibility of contamination from the air, and he acknowledged that, in fact, there was slight leakage of air during the length of the experiment. But Ramsay itemized the steps that he and Cameron took to prove that they had effectively isolated atmospheric contamination as the cause of the neon. On the anomalous neon, he stood firm. "The detection of neon," Ramsay wrote, "is open to no such objection." He discussed their efforts to avoid contamination with the neon and wrote that, after purifying the apparatus, the only contaminant they could detect was hydrogen:

> The vacuum tube had been previously run a great number of times at different pressures, washed out with air, and finally showed traces of hydrogen, and hydrogen alone. It is inconceivable that neon can have resulted thus by chance from two experiments with water, where in each case the residues were tested between those from numerous similar experiments with other solutions, in which no neon was detected. (Cameron and Ramsay, 1908, 997, Part III)

Two months later, in November 1908, Rutherford and Royds said that they attempted to replicate Ramsay's neon claim. They found no neon. Instead, they speculated about how Ramsay could have made a mistake. They wrote that, according to the information given in Ramsay's paper, Ramsay's detection limit for neon was questionable, and on this basis, they imputed that the neon claimed by Ramsay came from atmospheric contamination. (Rutherford and Royds, 1908)

In his 1913 book, Rutherford gave his opinion with greater certainty. "The neon and argon observed in the experiment of Cameron and Ramsay," Rutherford wrote, "were undoubtedly derived from the small quantity of air which admittedly leaked into the apparatus." (Rutherford, 1913, 321, Radioactive Substances) Despite Rutherford's conviction, he did not, however, support his speculation with evidence.

The Curie and Rutherford papers are the only ones I am aware of that made direct attempts to challenge the 1907-08 Ramsay claims. The mystery of whether Ramsay's neon was, in fact, from an air leak remains unsolved. Rutherford's thesis does not directly disprove Ramsay's neon. On the other hand, the lack of independent confirmation for Ramsay's claims works against Ramsay's claim.

Historical Accounts of Ramsay's Claims

Several historians have written about Ramsay's 1907-8 transmutation claims. I have examined the work of these authors:

William A. Tilden's book *Sir William Ramsay* (1918)
Benjamin Harrow's article in *The Scientific Monthly* (1919)
Alfred W. Stewart's book Recent Advances in Physical and
   Inorganic Chemistry (1920)
H. Stanley Redgrove's book *Alchemy: Ancient and Modern*
   (1922)
Morris W. Travers' book *A Life of Sir William Ramsay* (1956)
Thaddeus Trenn's article in *Ambix* (1974)
Milorad Mladjenovic's book The History of Early Nuclear
   Physics, 1896-1931 (1992)

None of their accounts provides any additional technical information about Ramsay's 1907 transmutation claim; however, a fascinating pattern emerged in my review of their accounts. I did not seek out authors who had any specific bias on the matter. Nevertheless, the first four — Tilden (1918), Harrow (1919), Stewart (1920), and Redgrove (1922) — all conditionally accepted Ramsay's claims. In later years, there was a dramatic shift. Travers (1956), Trenn (1974), and Mladjenovic (1992) were all dismissive of the claims. The odds of this pattern as a random coincidence are too great. What is the explanation for this?

I have seen no evidence that Travers, Trenn and Mladjenovic had access to different scientific information about the Ramsay transmutation claims than Tilden, Harrow, Stewart, and Redgrove. Nor have I seen evidence that the additional scientific knowledge available to the later authors contributed to more-precise or -insightful accounts of the Ramsay transmutation claims.

Here's what I do know. The later authors wrote their accounts after the development of high-energy physics in the 1930s and 1940s, after the acquisition of nuclear science by physicists, and after the bomb-making achievements of nuclear physicists in World War II. And there is one more point: In 1919, Rutherford tried to use "Radium C" to disintegrate matter. He succeeded, but he claimed only that he caused a partial disintegration of an atom; he made no claim about an elemental transmutation. He claimed that he caused a proton to be emitted from a nitrogen nucleus. Rutherford was well-liked and well-respected, and because he avoided making a claim of transmutation, his incremental claim of using an alpha particle to knock off a proton from nitrogen, even though that wasn't exactly what happened, was easily accepted.

News Media Accounts of Ramsay's Claims

Ramsay's 1907 transmutation claims got worldwide attention. The news broke on May 3, 1907, in the *Baltimore Sun*, and it was heralded as a GREAT DISCOVERY.

Of course, the *Baltimore Sun* was wrong. Ramsay claimed the production of lithium from copper sulfate, not the other way around. The *Sun* got the story backward. Ramsay had not made any presentation or published a paper, but he had sent a letter — which he requested remain private — to a friend and colleague, Ira Remsen, the president of Johns Hopkins University in Baltimore.

The day after the news broke, on May 4, 1907, Ramsay gave his first presentation on his transmutation claims. He presented his findings at the Chemical Society meeting in London.

The following day, on May 5, the *New York Times* ran a story that corrected the information reported in the *Baltimore Sun* story. The reporter for the *Times* had contacted Ramsay directly on May 4, and Ramsay vehemently denied the *Baltimore Sun* story, saying that it was "all a hoax." Ramsay denied knowledge of how the information was leaked to Johns Hopkins University.

Two news stories followed on July 27, 1907, which were guaranteed to aggravate Ramsay's scientific colleagues. There was no published paper from which they could analyze his claims or attempt a replication. Readers familiar with the Martin Fleischmann and Stanley Pons "fusion" announcement of 1989 will recognize the similar circumstance.

GREAT DISCOVERY

Hopkins University Learns Of Copper Made By A Remarkable Process.

SIR WM. RAMSAY DID IT

Radium Vapor Basis Of Find Of World Importance.

BALTIMORE MAY BE AFFECTED

Largest Copper-Smelting Industry On The Globe Exists Here--While The Latest Wonderful Attainment In Synthetic Chemistry Is Known In This City, London Will Not Hear It Until Tomorrow.

Headline from Baltimore Sun *May 3, 1907, news story*

NO SYNTHETIC COPPER.

Sir William Ramsay Says Report of His Discovery is a Hoax.
Special Cablegram.

Copyright, 1907, by THE NEW YORK TIMES CO.

LONDON, May 4.—"All a hoax," said Sir William Ramsay when I asked him to-day for particulars of his alleged discovery, reported in the Baltimore Sun and republished all over the world, of a synthetical process of making copper. The great scientist

Excerpt from New York Times *May 5, 1907, news story*

Leading the procession was the prestigious British medical journal *The Lancet*, with a short article called "Modern Alchemy: Transmutation Realized." It began with this statement:

> [Ramsay] has promised to communicate to the Chemical Society shortly a full account of his researchers on the radium emanation. The occasion will mark a great epoch in the history of chemical science since his investigations have shown that a given element under the powerful action of the radium emanation undergoes 'degradation' into another. In short, the transmutation of elements is actually *un fait accompli*.

Nothing could have been further from the truth; Ramsay had no published paper, no independent replication. *The Lancet* writers had stars in their eyes. They quickly jumped to the idea of transmutation of metals and ceremoniously exhumed alchemy from its burial: "These remarkable discoveries remind us again of the extraordinary prescience of the ancients, of the presentiments of the alchemists who evidently had some sort of a conviction that, after all, there is a primary matter from which all other elements are formed by various condensations."

The *New York Times* picked up the news from *The Lancet* and ran a follow-up story. The headline read "TURNS COPPER INTO LITHIUM: Sir William Ramsay Effects the Transmutation of Elements, Sought by the Alchemists." The *Times* article quoted heavily from *The Lancet's* article, including the dramatic statement about "an epoch in the history of chemical science," and said that it was *un fait accompli*. Certainly, this news reported in the *New York Times* was heard around the world.

The next day, July 28, 1907, the *New York Times* published four articles on the story. One was a short, moderately skeptical commentary on the Ramsay claims. Two were encouraging commentaries. The fourth was a personal interview Ramsay gave to a reporter for the *Times*. This new story — Ramsay's communication to the popular media, rather than the scientific media, about creating lithium — gives a possible clue to why Curie may have attempted to replicate Ramsay's minor claim of lithium rather than his major claim of neon.

Scientists already knew that argon-type gases, specifically helium, were produced by the disintegration of radioactive elements. The production of another gas, neon, would not have been earth-shattering news. Lithium, however, was a metal, and its creation would have been a revolutionary accomplishment for Ramsay.

TURNS COPPER INTO LITHIUM

Sir William Ramsay Effects the Transmutation of Elements, Sought by the Alchemists.

CHANGE CAUSED BY RADIUM

Possibility That Sodium and Potassium Are Also Formed.

GOLD MAY BE DECOMPOSED

Theory That the Elements of High Atomic Weight Will Disappear— —Statement by Ramsay.

Special Cablegram.

RAMSAY SURE OF DISCOVERY

Tells The New York Times That Repeated Experiments Were Made.

PROOF BY SPECTROSCOPE

Red Line of Lithium Appeared After Copper Had Been Treated.

IT DID NOT SHOW BEFORE

Sir William Says the Only Conclusion He Could Draw Was That the Copper Had Been Degraded.

Special Cablegram.

Headlines from New York Times *July 27 and 28, 1907, news stories*

On August 1, 1907, Ramsay gave another conference presentation, this time at the annual meeting of the British Association for the Advancement of Science. An article in *Nature* reviewing the meeting, published on Aug. 29, briefly mentioned Ramsay's set of claims. "The importance of [Ramsay's] discovery is that this is the first time the nature of the products of radioactive disintegration have been found to be controllable." (*Nature*, 1907, 457) None of his papers on this "discovery" had published yet. Two years later, on March 25, 1909, Ramsay claimed to transmute certain elements into carbon though he made no attempt to hide the fact that his data was marginal. (*New York Times*, March 26, 1909)

From Death to Disrepute

In 1912, Ramsay and a few other chemists reported on another set of transmutation experiments. The following year, he retired from University College. In 1916, after suffering from polyps in his nasal cavity for many years, Ramsay died from cancer.

The power of Rutherford's 1907 insinuation that Ramsay had made a sophomoric error by failing to properly eliminate neon contamination from air leaks marked Ramsay for life and beyond. Trenn, in collaboration with Rutherford's biographer Arthur Eve and Rutherford's colleague Bertram Borden Boltwood, left Ramsay's legacy in tatters.

"Ramsay was thus found to have made a blunder of the first magnitude." Trenn wrote. "He has accordingly been stigmatized as a great chemist who, [according to Rutherford's biographer Arthur Eve,] 'made incursions into radioactivity which were singularly unfortunate,' and ridiculed [by Bertram Borden Boltwood] for not also suggesting that 'radium emanation and kerosene [might] form lobster salad.'"

In the 1920s, Rutherford would gain credit and respect from the scientific community for his suggestion that radium emanation could, as Ramsay had claimed 13 years earlier, effect elemental change at the hand of man. The scientific community credited Rutherford for such an accomplishment, which, as I reveal in Chapter 22, actually belonged to his student.

CHAPTER 11

Soddy: Keen Observer and Visionary

Chemist Foresees the Potential of Nuclear Power and the Value of Transmutation (1908)

In August 1907, William Ramsay gave a paper on his transmutation claims at the annual meeting of the British Association for the Advancement of Science. Three months later, Frederick Soddy reported something peculiar that he and his colleague, Thomas D. MacKenzie, a Carnegie research scholar, had seen in one of their experiments. By this time, Soddy had also begun to see glimpses of what would later be known as nuclear science.

Mysterious Source of Gases

On Nov. 7, 1907, Soddy stood before his peers at the Chemical Society of London and reported his and MacKenzie's work. The objective of their experiment was only to determine the conductivity of certain gases, specifically, the noble gases: helium, neon, and argon. Soddy and MacKenzie were using a Crookes tube, the same kind of device used by Wilhelm Roentgen when he discovered X-rays. (Soddy and Mackenzie, 1908)

At one point while they were preparing to start an experiment, they checked to confirm that no rare gases were inside the tube, and to their amazement, they found some. They had just completed a process to empty the glass tube of any "occluded gases by heating and passing a discharge heavy enough to fuse the electrodes." They ran several other tests, checking for possible sources of contamination and influx but found no evidence of leaks.

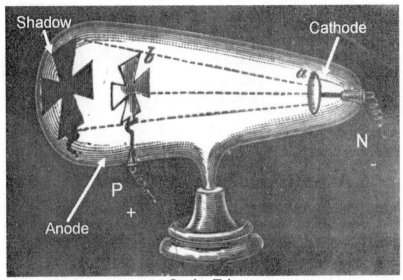

Crookes Tube

With no other explanations available to them, they concluded that the aluminum electrodes in the tubes must have contained helium, neon and argon gas before the experiments started. To them, it was nothing spectacular to report. They did not speculate on the possibility that the gases could have been formed by elemental transmutation.

There is not enough information to evaluate the alternate possibility in their experiment, that the gases were produced by transmutation. However, experiments that occurred later might shed some light. The first ones took place in 1912. (Chapter 13) Others took place in 2000, when a scientist at the SRI International laboratory in Menlo Park, California, claimed that helium was being occluded in palladium rather than being generated on the surface of the metal.

At least in the case of helium, the rare gas does not permeate or diffuse through intact and defect-free metals, and thus cannot be easily occluded or, if somehow trapped inside, released from bulk metal.

Visions of a Nuclear Future

In 1908, Soddy gave a series of lectures at the University of Glasgow, where he was a lecturer in physical chemistry and radioactivity. The following year, he published the book *Interpretation of Radium*, based on those lectures. His interpretation not only of radium but also of radioactivity led him to an extraordinary, if not Utopian, vision of the potential of nuclear science:

> A race that could transmute matter would have little need to earn its bread by the sweat of its brow. If we can judge from what our engineers accomplish with their comparatively restricted supplies of energy, such a race could transform a desert continent, thaw the frozen poles and make the whole world one smiling garden of Eden. Possibly they could explore the outer realms of space, emigrating to more favorable worlds as a superfluous [people] today emigrate to more favorable continents. (Soddy, 1909, 244)

Soddy was not blind to the potential use of atomic energy (later called nuclear energy) for destruction. He recognized that the new field of radioactivity (later known as nuclear science) did not belong exclusively to the domain of either chemistry or physics. In the first chapter of his book, he explained the impossibility of providing a simple analogy to explain the new science to the lay public:

> Radioactivity is a new primary science owing allegiance to neither physics nor chemistry, [and] as these sciences were understood before [radioactivity's] advent, ... the old laws of physics and chemistry concerned, almost wholly with external relationships, do not suffice. ... Is it possible to give, by the help of an analogy to familiar phenomena, any correct idea of the nature of this new phenomenon "radioactivity"? The answer may surprise those who hold to the adage that there is nothing new under the sun. Frankly, it is not

possible, because in these latest developments science has broken fundamentally new ground, and has delved one distinct step further down into the foundations of knowledge. (Soddy, 1909, 3)

Had anyone twelve years ago ventured to predict [the transmutation of] radium, he would have been told simply that such a thing was not only wildly improbable, but actually opposed to all the established principles of the science of matter and energy. So drastic an innovation was, it is true, unanticipated. Radium, however, is an undisputed fact today, and there is no question, had its existence conflicted with the established principles of science, which would have triumphed in the conflict. Natural conservatism and dislike of innovation appear in the ranks of science more strongly than most people are aware. Indeed, science is no exception. (Soddy, 1909, 4)

Soddy's observation about conservatism in science and, specifically, the fact that most people are generally unaware of this conservatism is a concept of immense significance today, just as it was then. The force of conservatism in science opposes the force of innovation and discovery in science, and this will become clearer in the next few chapters.

The Power of Transmutation

In 1909, nuclear fission, the process used by nuclear power plants to make heat and electricity for much of the industrialized world, had not been discovered. But Soddy recognized its potential then. In 1938, Otto Hahn and Fritz Strassmann demonstrated nuclear fission, and the next year, Lise Meitner and Otto Frisch explained the process. Thanks in part to the 1903 work of Marie and Pierre Curie and Gustave Bémont, who recognized the heat released by radium, as well as by Einstein's 1905 theory of special relativity, Soddy was quick to recognize the potential for this new class of transmutation-based (nuclear) energy production, one that did not rely on combustion but used the immense energy densities that only nuclear reactions can provide, as he wrote in *Interpretation of Radium:*

> I have already referred to the total amount of energy evolved by radium during the course of its complete change. It is about two million times as much energy as is evolved from the same weight of coal in burning. The energy evolved from uranium would be some fourteen percent greater than from the same weight of radium. This bottle contains about one pound of uranium oxide, and therefore about fourteen ounces of uranium. Its value is about £1. Is it not wonderful to reflect that in this little bottle there lies asleep and waiting to be evolved the energy of about 900 tons of coal? The energy in a ton of uranium would be sufficient to light London for a year. The store of energy in uranium would be worth a thousand times as much as the uranium itself, if only it were under our control and could be harnessed to do the world's work in the same way as the stored energy in coal has been harnessed and controlled.
>
> There is, it is true, plenty of energy in the world which is practically valueless. The energy of the tides and of the waste heat from steam falls into this category as useless and low-grade energy. But the internal energy of uranium is not of this kind. The difficulty is of quite another character. As we have seen, we cannot yet artificially accelerate or influence the rate of disintegration of an element, and therefore the energy in uranium, which requires a thousand million years to be evolved, is practically valueless. On the other hand, to increase the natural rate, and to break down uranium or any other element artificially, is simply transmutation. If we could accomplish the one, so we could the other. These two great problems, at once the oldest and the newest in science, are one. Transmutation of the elements carries with it the power to unlock the internal energy of matter, and the unlocking of the internal stores of energy in matter would, strangely enough, be infinitely the

most important and valuable consequence of transmutation. (Soddy, 1909, 229)

More Valuable Than Gold

Two years later, on Feb. 19, 1911, the *New York Times* published "ALCHEMY, LONG SCOFFED AT, TURNS OUT TO BE TRUE: Transmutation of Metals, the Principle of the Philosopher's Stone, Accomplished in the 20th Century." In the article, the author quoted extensively from Soddy's book, wrote about the discoveries of natural transmutations and attempted to use an analogy:

> Hitherto, chemists have dealt with atoms as architects have dealt with bricks. Bricks have been built up into different forms, and produced widely different results, but they were always bricks. Nobody ever got [around] that fact. What would be the surprise of a builder if he suddenly discovered that a brick would pull a wagon or keep him warm, or go off with a bang like gunpowder? His surprise would be no greater than that of the chemist when he discovered that atoms of matter, of their own accord, were prepared to do things quite as startling.
>
> Since experiments have proved that elements believed to be totally different can change one into the other, and since these elements are quite common and ordinary, though they have been found to behave in a strange way, there is no reason to believe that they are exceptional. Uranium changes into radium: part of radium becomes a gaseous body which in no time at all will turn back again into a solid, and then go on changing from one thing to another, tending toward something that is very probably just common lead, though this last [step] has not been proved. Moreover, radium under different circumstances turns into different things. If this is not transmutation, what is? Science is on the verge of an epoch-making discovery that may utterly revolutionize

not only our ideas of chemistry, but, perhaps, of astronomy and geology, too, and, what is more important, may put in the hand of man the power to harness nature as was never dreamed of before, and change civilization to a fairy tale. (*New York Times, Feb. 19, 1911*)

A century later, the phenomenon of disintegration, now called radioactive decay, and processes such as fission and fusion, all of which involve the transmutation of elements, are accepted in science without hesitation. The science of radioactivity, as it was called in 1911, now contributes to a foundational understanding of physics and chemistry, not to mention a multitude of technologies that have evolved from it. The allusion to the dreaded alchemy did not escape the *Times'* author's attention, but the author realized that something else was more important than the alchemists' dream of producing gold:

This is the alchemy of the 20th Century. With all this mystery and topsy-turvyiness it is easy to see how poor a chance the alchemists have of ever reaching a simple method of making gold. Probably in the course of trillions and trillions of years something is turning to gold. But what? And even if the [correct starting] substance was discovered and the change could be chemically made, the energy that would be given off in bringing about [the transmutation] would be sufficient to run the world's business for it for many a year. The value of the discovery would not lie in the gold produced.

Just here comes the wonder that opens before the world with this discovery of the energy that is stored in a harmless looking substance like radium. Presumably there is equal energy and greater, in every ordinary substance about us, if we could get it out. The axles of carriage wheels could run the vehicle faster than any automobile; the iron in the grate could give out heat and to spare, without a particle of coal.

Frederick Soddy

Ahead of His Time

Two years later, in 1913, Soddy realized something that now comprises another fundamental concept of nuclear chemistry and nuclear physics. As noted earlier, he developed the concept of isotopes: atoms that have the same nuclear charge but different mass. (Soddy, 1913) For this, he was awarded the 1921 Nobel Prize in chemistry.

Soddy struggled with his many interests within and outside of science. At the Oxford University, he tried to introduce reforms in the methods of teaching physical chemistry, but he was dissatisfied with the slow progress of change.

Outside of science, he was an advocate for Irish independence and voting rights for women. He also had a strong interest in economics. Eventually, votes were given to women and independence to Ireland, but because Soddy was ahead of his time, he experienced frustration with his advocacy. (Paneth, 1964, 88)

Failure of Imagination

Soddy's foresight was not appreciated by all of his peers. Caltech physicist Robert Andrews Millikan (1868-1953), in his book *Science and the New Civilization*, had this to say about visionaries like Soddy:

> Under the stimulus of the discovery of the enormous quantities of energy evolved in the disintegration of uranium and thorium, we have often imagined, and sometimes incautiously stated, that there might be similar amounts of available energy locked up in the common elements, releasable, perchance, by getting them to disintegrate, as uranium and thorium spontaneously are doing. And engineers, physicists and laymen alike have talked glibly about "utilizing this source of energy when the coal is gone." (Milliken, 1930, 94-5) ...
>
> If Einstein's equation and Aston's curve are even roughly correct, as I'm sure they are, for Dr. Cameron and I

have computed with their aid the maximum energy evolved in radioactive change and found it to check well with observation, then this supposition of an energy evolution through the disintegration of the common elements is from the one point of view a childish Utopian dream, and from the other a foolish bugaboo. (Millikan, 1930, 95) ...

The energy available to him through the disintegration of radioactive, or any other, atoms may perhaps be sufficient to keep the corner peanut and popcorn man going, on a few street corners in our larger towns, for a long time yet to come, but that is all. (Millikan, 1930, 111) ...

The energy available to him through the building-up of the common elements out of the enormous quantities of hydrogen existing in the waters of the earth would be practically unlimited provided such atom-building processes could be made to take place on the earth. But the indications of the cosmic rays are that these atom-building processes can take place only under the conditions of temperature and pressure existing in interstellar space. Hence, there is not even a remote likelihood that man can ever tap this source of energy at all. The hydrogen of the oceans is not likely to ever be converted by man into helium, oxygen, silicon or iron. (Millikan, 1930, 111)

Sir Arthur C. Clarke, one of the world's greatest visionaries and futurists, was familiar with this mode of thinking. He distilled it in his 1962 book *Profiles of the Future*: "With monotonous regularity, apparently competent men have laid down the law about what is technically possible or impossible – and have been proved utterly wrong, sometimes while the ink was scarcely dry from their pens. On careful analysis, it appears that these debacles fall into two classes, which I will call Failures of Nerve and Failures of Imagination."

Clarke's "First Law," has proved true time and time again: "When a distinguished but elderly scientist says something is possible, (s)he is almost certainly right. But when (s)he says something is impossible, (s)he is very probably wrong."

CHAPTER 12

Visionary Scientists — or Not?

Ramsay Foresees Coal Decline; Rutherford Fails to See Atomic Energy

The previous chapter revealed how Frederick Soddy envisioned the potential of radioactivity for new energy sources and applications. Ramsay, too, understood the energy of the atom and its possible application to future technology. Ramsay was also concerned about England's heavy reliance on coal, a nonrenewable energy source.

Ramsay's Visionary Coal Prediction

According to a Sept. 11, 1911, article in the *New York Times*, Ramsay spoke at the British Association for the Advancement of Science, expressing his concerns about the future of energy in the country. He made a surprisingly accurate prediction. "In [Sir William Ramsay's] very calm and careful but significant address at the recent meeting of the British Association," the *Times* wrote, "[he] declared that the coal fields of the United Kingdom will be exhausted in 175 years."

Ramsay was correct, although the United Kingdom depleted its coal in half the time he predicted. Technically speaking, fossil fuels may never be exhausted. As scarcity sets in, however, market prices will adjust accordingly.

His prediction was surprisingly accurate, considering that he made it at the very peak of U.K. coal production and that no downward trend data existed. The graph below is based on data from the U.K. Department of Energy and Climate Change and proves that Ramsay was qualitatively correct.

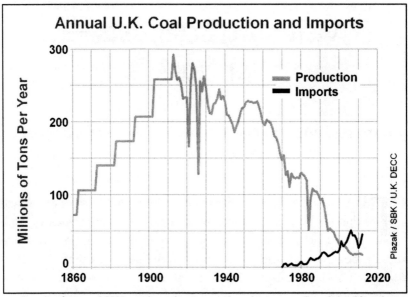

Graph of annual U.K. coal production and coal imports. Graph by Plazak, data from U.K. Department of Energy and Climate Change

The Sept. 19, 1911, *New York Times* article "England Without Coal" was no exaggeration. Ramsay saw the long view, and the *Times* writer understood that nothing else could be as debilitating to the country as its loss of coal without a suitable replacement. Ramsay and other scientists of the day knew the potential of uranium, but not until several decades later did scientists find a way to harness nuclear fission energy. The *Times*' article gave a snapshot of the energy outlook a century ago:

> Sir William Ramsay [does not] hold out any specific hope that by the time the coal of the kingdom is gone, any other source of energy can adequately replace it. ...
> Obviously there is no practical relief in this direction for

the disaster he describes as impending. He can only suggest partial measures — the more earnest study of pure science, more thorough training in technical science [and the] restriction of waste. ...

Other scientists however, are not so gloomy in their forecast. It is pointed out that there are, in other lands, constantly more accessible large supplies of coal, and it is suggested that in the next two centuries, England, with its admirable talent for organized enterprise, will be able to get its full share of these supplies. ... Engineers declare, for instance, that those known to exist on the Mexican East Coast from Vera Cruz northward are 10 times greater than those of the famous Baku region. ...

Something may turn up, but it is discouraging to hear from a great scientist that there is nothing of which he can speak with any confidence. England without coal — or its equivalent — would face bankruptcy and decay. The only consolation is that the scientists of two centuries ago could not speak with any confidence of the mighty energy of steam and electricity that has since been subdued to the uses of the world. (*New York Times*, 1911)

The concept of England obtaining "its full share" of fossil fuels from Mexico is profound in that it foreshadowed a growing undercurrent that, by now, is a dominant component of geopolitical agendas. A graph by Roger Fouquet illustrates the shift of energy sources in the United Kingdom, beginning 500 years ago, as it transitioned into and out of fossil fuels. (Fouquet, 2010)

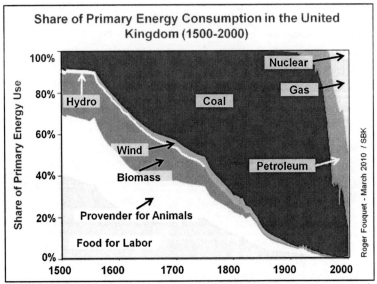

500-year shift of energy sources in the United Kingdom (Fouquet, 2010)

Rutherford's Not-So-Visionary Prediction

Rutherford is famous for his botched prediction of nuclear energy, which he made while giving a talk at the British Association for the Advancement of Science on Sept. 11, 1933.

Newspapers across the country picked up an Associated Press story quoting him. "The energy produced by the breaking down of the atom is a very poor kind of thing," Rutherford said. "Anyone who expects a source of power from the transformation of these items is talking moonshine."

The *Times of London* mixed in Rutherford's language with its reporters' own language in its article on Rutherford's presentation. "We might," the *Times* wrote, "in these processes, obtain very much more energy than the proton supplied, but on the average we could not expect to obtain energy in this way. It was a very poor and inefficient way of producing energy, and anyone who looked for a source of power in the transformation of the atoms was talking moonshine. But the subject was scientifically interesting because it gave insight into the atoms." (*Times of London*, 1933)

Power Source Not in Atoms

LEICESTER, England, Sept. 11 (P).—Lord Rutherland, at whose Cambridge laboratories atoms have been bombarded and split into fragments, told an audience of scientists today that the idea of releasing tremendous power from within the atom is absurd.

He addressed the British Association for the Advancement of Science in the same hall where the late Lord Kelvin asserted twenty-six years ago that the atom is indestructible.

Describing the shattering of atoms by use of 5,000,000 volts of electricity, Lord Rutherford discounted hopes advanced by some scientists that profitable power could be thus extracted.

"The energy produced by the breaking down of the atom is a very poor kind of thing," he said. "Anyone who expects to source of power from the transformation of these atoms is talking moonshine.

"We hope in the next few years to get some idea of what these atoms are, how they are made and the way they are worked," he continued.

CHICAGO, Sept. 11 (P).— Chemists attending the American chemical society meetings today agreed with Lord Rutherford that no great power source is to be expected in unlocking atomic energy.

HOPE OF TRANSFORMING ANY ATOM

What, Lord Rutherford asked in conclusion, were the prospects 20 or 30 years ahead?

High voltages of the order of millions of volts would probably be unnecessary as a means of accelerating the bombarding particles. Transformations might be effected with 30,000 or 70,000 volts. Two methods of attack would be open: one by intensive streams of particles of relatively low voltages, the other by the use of higher voltages up to 5,000,000 volts, which ought to be sufficient to break down any atom on this earth, and he believed that we should be able to transform all the elements ultimately.

We might in these processes obtain very much more energy than the proton supplied, but on the average we could not expect to obtain energy in this way. It was a very poor and inefficient way of producing energy, and anyone who looked for a source of power in the transformation of the atoms was talking moonshine. But the subject was scientifically interesting because it gave insight into the atoms.

Those processes had been going on for thousands of millions of years in our sun and in other hot stars, and the whole problem of the abundance of the different types of the elements, therefore, was probably decided by those processes of the building up and destruction of atoms, due to the emission of particles at enormous pressures and temperatures. He need hardly say that the subject of atomic transformation was the great subject of physics to-day, and there was an ever-increasing attack on this most formidable of problems because it was hoped that we might by solving it get some idea of the way the nuclei of atoms were made and of the way in which the atoms disintegrated.

September 12, 1933, Scranton Republican *story about Rutherford's view on the prospects of atomic energy (left); September 12, 1933,* Times of London *story about Rutherford's view on the prospects of atomic energy (right)*

Rutherford's perspective was not limited to his moonshine comment in 1933. His attitude had been consistent for a decade, as evident in his 1924 article "The Energy in the Atom. Can Man Utilize It?" Rutherford wrote the article while he was the president of the British Association for the Advancement of Science. In fact, despite Einstein's 1905 theory of special relativity, Rutherford — as a spokesman for British science — suggested that the prospects of unleashing the power of the atom had diminished by 1924:

> Unfortunately, although many experiments have been tried, there is no evidence that the rate of disintegration of these [radioactive] elements can be altered in the slightest degree by the most powerful laboratory agencies. With increase in our knowledge of atomic structure, there has been a gradual change of our point of view on this important question, and there is by no means the same certainty today as a decade ago that the atoms of an element contain hidden stores of energy. (Rutherford, 1926, 109-11)

Clearly, Rutherford was not motivated to pursue nuclear transmutations for a possible new source of energy. Ramsay and other chemists, however, were.

CHAPTER 13

Three Mysterious Rare Gases

Finally, Man-Made Nuclear Transmutations, but Physicists Are Not Convinced (1908-1914)

By this time, scientists knew that elemental transmutations were directly related to the prospect of atomic energy. Until now, no scientist had performed a confirmed man-made transmutation from a stable element. In a scientific paper that appeared several years later, German chemist Fritz Paneth summarized the major distinctions between the concepts of natural versus man-made transmutation:

> [Thus far,] man has no power to influence [transmutation]. The production of helium from radium takes place with absolute constancy, and no means at man's disposal, neither extremely high nor extremely low temperature, nor very high nor very low pressure, nor electric or chemical energy, can quicken or retard the rate of the transmutation. (Paneth, 1926)

Between 1912 and 1914, half a dozen scientists attempted man-made transmutations using low-energy methods rather than high-energy physics: They passed high-voltage electrical discharges in vacuum or through hydrogen at low pressures in cathode-ray tubes. These experiments were the first attempts at man-made transmutation without the use of alpha bombardment. They were the first confirmed man-made transmutations from stable elements.

Paneth had dismissed the 1912-1914 transmutations without performing a careful scientific analysis, and he claimed that these results were all erroneous. Paneth, however, was tainted by self-interest. At the time he wrote the statement above, he also claimed that he and his colleague Kurt Peters were the first scientists to perform man-made transmutations.

The 1910s experimenters ran continuous, high-voltage electric discharges through the tubes for hours or days, then analyzed the residual gases with optical emission spectroscopy.

The same electrical current used for the discharges was used for the analysis. The discharge excited the contents of the tube. The excitation, in turn, caused the emission of light, which was directed to a prism or a diffraction grating to reveal the spectrum. The lines on the spectrum provided a fingerprint, or signature, of the constituent elements. The spectrum was captured on photographic film or observed optically. Spectroscopy had been well-established by this time for such qualitative analysis.

In 1907, William Ramsay claimed he had made the first man-made transmutation from a stable element. Ramsay used radium as his source of alpha particles to bombard his target materials. This was front-page news at the time, but his results were never replicated. Although science experiments are occasionally accepted before being replicated, this one was far too radical to be accepted by the scientific community without independent confirmation.

In 1907, something else interesting happened. Frederick Soddy and his colleague Thomas D. MacKenzie, observed, to their great surprise, that two rare gases, helium and neon, appeared in one of their experiments. They were preparing to run tests with cathode-ray tubes. Their plan was to fill the tubes with rare gases, pass a high-voltage electric discharge through them, and thus determine the electrical conductivity of the gases.

But as they were preparing for their experiment, they cleaned their cathode-ray tube — they thought — by running an electric discharge through it. They hadn't put any helium or neon gas into the tube. After "cleaning," they double-checked to make sure that the tube was clean. It wasn't: They saw the spectra of helium and neon!

They had no idea where the gases came from, but they assumed there was a reasonable explanation. Rather than making a claim of transmutation, they said that the gases had been occluded, that the rare gases had been hiding out deep in the metal cathodes. They explained their unspectacular results as such, and no one gave them a second thought. Well, almost no one.

Sometime after 1907, Ramsay had to return the valuable radium, which he had borrowed from the Radium Institute of Vienna. He began wondering whether beta particles (electrons) might also be able to induce transmutations, just as alpha particles were then known to do. He, as he wrote, in conjunction with his colleague John Norman Collie (1859-1942), repeated Soddy and MacKenzie's experiment and replicated their results three times. Ramsay and Collie, too, got neon and helium. Ramsay reported this in 1912 in a short letter to *Nature*. (Ramsay, 1912)

A year later, at a London scientific meeting, Ramsay announced more details about his replication of Soddy and MacKenzie's work. Collie was not listed as an author on Ramsay's paper. (Ramsay 1913) Instead, Collie and another chemist, Hubert Sutton Patterson, reported their own experiments.

That same week, Joseph John Thomson, a physicist at the Cavendish laboratory of the University of Cambridge said that he, too, replicated the experiment and got helium and neon.

Thomson announced his replication; however, he claimed the gases weren't from transmutation but had been hiding out, occluded in the metal electrodes, just as Soddy and MacKenzie had suggested.

But here's the real kicker: Thomson said that he also got a third gas, which we now know is tritium. Without intending to and without making a claim, Thomson confirmed low-energy elemental transmutations in 1913.

Not only did Thomson do a poor job of supporting this half-baked occlusion hypothesis, the following year Collie, Patterson and a third chemist reported a rigorous battery of tests to isolate the possibility of both occluded gases and leaks from the atmosphere. The chemists defended their claims; they disproved the occlusion hypothesis.

Gas Through Metals

Before we discuss the 1912-14 research, it would be helpful to summarize the characteristics of helium permeation (or solubility) in metals as well as in glass. In short, helium does permeate glass, at room temperature and pressure, but only very slowly. It permeates much faster with elevated temperatures and/or pressures.

On the other hand, helium does not, in most cases, permeate metal. Some people confuse the behavior of helium in hydride-forming metals with the behavior of hydrogen in hydride-forming metals. The two gases, despite being neighbors on the periodic table, behave very differently in such metals.

In 2010, I performed a literature search of the behavior of hydrogen and helium in metals for a news investigation. The following table shows a summary of my findings. (See Appendix B for references.)

Gas Behavior in Hydride-Forming Metals at or Near Standard Temperature and Pressure	Hydrogen	Helium
Readily permeates hydride-forming metals	Yes	No
Diffuses through defects, cracks or grain boundaries in metals	Yes	Yes
Soluble (dissolves) in hydride-forming metals	Yes	No

Helium Permeating Glass?

Glass is permeable to some gases, particularly to helium and slightly less so to neon. But permeation in glass occurs slowly. In addition to time, temperature and pressure, other factors that affect permeation include surface area, glass thickness, and glass type. The researchers in the 1912-14 period didn't know all this, but they tested for helium permeation in glass. During their experiments, they found no passage of helium to their detection limit.

As shown in the book *Fusion Fiasco: Explorations in Nuclear Research, Vol. 2*, many researchers, beginning in 1989, attempted experiments to create nuclear reactions by chemistry. Melvin Miles, an electrochemist at the U.S. Navy China Lake laboratory, was one of them.

The graph below shows how slowly helium permeates glass. In 1992, Miles performed electrolytic experiments for several weeks in search of the production of helium. Toward the end of the experiments, Miles collected the evolved gases. But before doing so, he flushed the flasks with boiled-off nitrogen, to make sure the flasks were free of helium. He collected the gases in a few hours and sent them off to a lab for analysis.

At the lab, under standard temperature and pressure, atmospheric helium was allowed to permeate the three Pyrex glass flasks for 100 days.

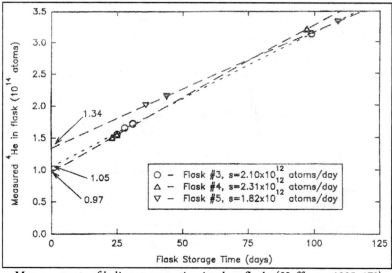

Measurements of helium permeation in glass flasks (Hoffman, 1995, 179)

Miles explained the data to me in an e-mail:

> After collection, the flasks were sealed, removed, and shipped to Rockwell International for the helium measurements. All helium measurements were performed by Rockwell, which produced the graph. The time scale in the graph starts on the day the gas collection was completed. They made the first measurement of flask #3 at day 25, #4 at day 23 and #5 at day 36, and additional measurements out to

100 days. Rockwell extrapolated the curves back in time to the day (time zero) when I collected the gases. Those helium values represent the amount of helium produced in my experiments.

Miles' data shows that, over 100 days, helium permeated the glass and increased the concentration in the flasks — but very slowly.

Helium and Neon From Air Leakage?

The researchers in the 1910s also reasoned that the helium was not from air leaks, because they didn't detect nitrogen or oxygen, which, had there been a leak, would have been present along with helium.

Below is a table that shows the fractional concentrations in air by volume. The ratio between the most common constituents of air, nitrogen and oxygen, and the rare elements observed, helium and neon, is between four and five orders of magnitude. Any spectral lines of nitrogen or oxygen from an air leak that would have produced detectable helium would have been obvious.

The spectroscopy measurements at the time were primarily qualitative (identifying what), not quantitative (identifying how much), and determining the detection limits back then is difficult. For the early 1990s researchers, detecting helium inside glass discharge tubes required concentrations of rare gases much higher than those present in air.

Fractional Concentrations of Dry Air by Volume	
Substance	**% by Volume**
Nitrogen	78.08
Oxygen	20.95
Argon	0.93
Neon	0.0018 (18 ppm)
Helium	0.00052 (5 ppm)
Methane	0.0002 (2 ppm)
Krypton	0.00011 (1 ppm)
Hydrogen	0.00005 (0.5 ppm)

Electrons in Motion

By 1912, William Ramsay had attempted transmutation by bombardment with beta particles — in other words, electrons, rather than alpha particles. (*Nature*, 1913, 653) Ramsay thought he had succeeded. On July 18, 1912, *Nature* published a letter from Ramsay about experiments in which he claimed to produce helium and a trace of neon with betas instead of alphas. His brief letter wasn't a scientific paper, and it had little in the way of experimental details or data. However, he had broken the ice: He claimed that it was possible, with the emission of electrons, to induce man-made elemental transmutations. But this was difficult for the scientific community to accept. His suggestion ran contrary to the understanding of the fundamental forces of physics at the time. Scientists thought that electrons emitted by cathode-ray tubes didn't provide sufficient high-energy stimuli required to cause nuclear reactions.

"From these experiments," Ramsay wrote, "it would appear that not merely atoms of helium in rapid motion [alpha particles] are capable of communicating sufficient energy to molecules and atoms on which they impinge to cause them to disintegrate, but that electrons in motion, in the form of cathode rays, can be made to play a similar part." (Ramsay, 1912, 502) A few days later, the *New York Times* published the news.

> **HELIUM'S POWER ON ATOMS.**
>
> Sir William Ramsay Makes Some Notable Experiments.
>
> Special Cable to THE NEW YORK TIMES.
>
> LONDON, July 20. — Sir William Ramsay contributes to Nature details of experiments which he carried out with Prof. Norman Collie, which show that not merely are atoms of helium in rapid motion capable of communicating sufficient energy to molecules and atoms on which they impinge to cause them to disintegrate, but that electrons in motion in the form of cathode rays can be made to play a similar part.

The New York Times *reported on Ramsay's letter to* Nature

Taking the Lead

At the Chemical Society meeting in London on Feb. 6, 1913, Ramsay spoke about his findings. Collie and Patterson reported their results, too. They each reported that they had independently produced neon and helium in glass cathode-ray tubes filled with hydrogen exposed to high-voltage electric discharges. (*Nature*, 1913) Ramsay, despite his informal letter to *Nature* the year before, had not published his and Collie's results in 1912. Collie and Patterson were not claiming a replication of Ramsay's work at the 1913 meeting; they were staking their own claim for priority. Within a year, scientists published a dozen papers on this topic. Appendix C presents a list of those papers.

John Norman Collie

Collie and Patterson had started their own work separately, but when they realized they were working on the same problem from different points of view, they teamed up. Collie was a professor of organic chemistry at University College, London, and in 1913, he became the dean of Chemistry. He was an avid mountain climber, and a river and a town were named after him.

The University College, London, Web site says that Collie was known best for two scientific achievements. "He was the first [person] in the U.K. to use X-rays for medical purposes," the Web site says. "The second, more dubious, achievement was the first use of neon discharge tubes for display purposes."

Collie sitting next to an illuminated neon discharge tube, glowing red
Image: University College London

Neon had been discovered by Ramsay, who was also a professor at University College, London. Patterson was a professor at the University of Leeds. I've not been able to find out anything else about him, and the press liaison at the University of Leeds didn't know anything else, either.

Collie had come into this general line of work years earlier. He had noticed that certain minerals, such as sodalite and fluorspar, changed color when they were bombarded by cathode rays. At the same time, they produced a lot of gas. With the impetus from his work with Ramsay in 1912, Collie began a new set of experiments with cathode-ray tubes to figure out the nature of these gases.

A Blaze of Helium

Collie and Patterson performed related experiments with minor variations in apparatus and protocol. In one set of experiments, they developed a tube-within-a-tube configuration to minimize the possibility of helium or neon permeation from the atmosphere.

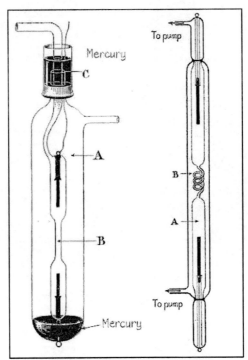

Double-chamber tubes used for cathode-ray experiments by John Norman Collie (left) and Collie and Hubert Sutton Patterson (right)

At times, they filled the space between the tubes with specific gases to test for diffusion. They wanted to check the degree to which helium or neon leaked through the glass. At other times, they placed the middle area under vacuum. Their experiments generally ran for three to six hours, rather than days, and under these conditions, they did not observe helium and neon permeation.

Collie and Patterson presented their research to the Chemical Society in a meeting room at Burlington House in London. A reporter from the *Morning Post* was there and recorded his account of the presentations.

> "[Collie] drew attention to the fact that Mr. Patterson, in [electrically] sparking the hydrogen [to observe the spectrum], had not [observed] helium, but neon. [Collie] had, he said, criticized Mr. Patterson's method of preparing hydrogen, and to avoid [a] possibility of error, Mr. Patterson had started by filling his apparatus with pure oxygen, but [after] pumping [the oxygen] out [he] got the same result.
>
> Another possibility had then suggested itself. While neon did not enter glass under ordinary conditions, might it not do so under the influence of the X-Ray discharge? To make certain on this point, the experiment tube was surrounded with another tube containing neon, and about the same result was obtained as with ordinary air. A similar experiment was made with helium with negative results.
>
> Last, before sending in his paper the previous week, [Collie] had used the outside vessel as a vacuum [a stronger vacuum than in an X-ray tube], and still the neon appeared, the quantity thus obtained being comparable with that present in about 2 cm^3 of air (Loud cheers).
>
> Last Friday and Saturday, he repeated, he had performed the experiment twice with the experiment tube surrounded by a vacuum. He had then asked himself whether there was anything else he could test. He had nothing [in] particular to do that afternoon, and he had decided to test whether there was anything in the outer chamber. He let a cubic centimeter of pure oxygen into the outer chamber, and he obtained a slight explosion due to hydrogen. He absorbed the oxygen in the usual way with carbon, but there was still some left, which he regarded as rather a nuisance. He repeated the process of absorption, but the gas still remained in relatively large amounts.

He decided to test it, and turned on the coil. The sight he then saw astounded him, for the tube was a blaze of helium, with some neon mixed. (Loud cheers). He communicated with Mr. Patterson, who repeated the experiment. Mr. Patterson at first found helium. Then [Patterson] put oxygen into the outer tube, and he found, instead of helium, a large quantity of neon ... (Cheers). (*Morning Post*, 1913)

Photograph of meeting room of the Chemical Society in Burlington House.

Collie performed 35 experiments and found neon in greater or less quantity every time. Patterson found neon in 15 experiments and, in four of them, some helium, too. They each performed a variety of blank (control) experiments and found no helium or neon.

Masson Replication

James Irvine Orme Masson (1887-1962) replicated one of the Collie-Patterson configurations and obtained neon. Testing for contamination, he found no other gas, which would have revealed an air leak, except a small amount of carbon monoxide. (Masson, 1913, 233)

Masson was born in Australia. He earned a degree in chemistry at Melbourne University in 1908 and continued his studies at University College, London, in 1910. He stayed there and became a professor of chemistry.

Drawing of 1886 Chemical Society meeting in Burlington House.

Technical Analysis of Experiments

I've been able to find only one historian who seriously evaluated the Collie/Patterson experiments. Alfred W. Stewart, professor of chemistry at Queen's University of Belfast and author of *Recent Advances in Physical and Inorganic Chemistry*, discussed their work in the 4th edition of his book, published in 1920. (Stewart, 1920, 193-6) A few years earlier, in his 1914 book *Chemistry and Its Borderland*, Stewart explained additional details about the experiment and carefully attempted to analyze for possible alternate explanations besides transmutation.

Stewart considered air leaks and ruled them out because nitrogen had not been detected. He considered traces of neon mixed with the oxygen and hydrogen gases and ruled them out based on blank (control) tests Collie and Patterson reported. He reviewed the researchers' tests to check for neon permeation and ruled out that possibility. He considered the idea that neon might be occluded in the electrodes but ruled that out because Collie and Patterson reported observing neon in some experiments without electrodes. Instead, they applied electric forces from a spiral-wound wire around the outside of the evacuated vessel.

Stewart recognized Masson's confirmation: "Masson repeated the experiments in accordance with their directions; and his results completely confirm the work of the two discoverers."

After a long list of alternative explanations, Stewart wrote that the only remaining conclusion "is to assume that the neon has in some way been produced within the tube." Here is his conclusion:

> On the evidence before us, it appears impossible to find any other solution of the problem. Two absolutely independent investigators, working without knowledge of each other's researches, have arrived at identical results.
>
> A third person has repeated the experiments and has confirmed the data fully.
>
> Evidently there can be no question as to the facts. Time will show whether any source of the neon has been overlooked; but at the present time it seems difficult to refuse to believe that either the hydrogen may have been transmuted into the elements of the inert gas series, or that the cathode discharge, acting on the materials of the glass vessels used, may have transmuted them instead of the hydrogen. (Stewart, 1914, 208-211)

Collie and Patterson's conclusion is worthy of display:

> *Conclusion.*
>
> Whatever the explanation of the above results may be, the facts seem to be the following:
> 1. Neon cannot be obtained from either the glass or the electrodes by heating alone.
> 2. The neon found is not due to air leaking into the pump or the apparatus during the experiment.
> 3. Glass, neither when heated to near its softening point, nor under the action of cathode rays, is permeable to ordinary neon or helium.
> 4. The hydrogen and the oxygen used in the experiments did not contain neon.
>
> Moreover, the appearance of helium and neon in the outer tube as well as neon in the inner tube is most important. It would appear impossible that particles of these gases are shot through the glass from the inner to the outer tube; the question obviously is where do these gases come from? The answer to that question cannot at present be given with any certainty, but it is hoped that this line of research may ultimately throw more light on the subject. Since this paper was read Sir J. J. Thomson, in a letter to *Nature* (**90**, 645) gives an account of some experiments he has been making that have a direct bearing on the phenomena in question.
>
> UNIVERSITY COLLEGE,   UNIVERSITY,
> LONDON.   LEEDS.

A Mysterious Gas

Physicist Joseph John Thomson (1856-1940), the Cavendish Professor of Physics at the University of Cambridge, sought to challenge the chemists' transmutation claims. Thomson was the expert on electrons; he discovered them using cathode-ray tubes. This earned him a Nobel Prize in physics in 1906. One of his other notable achievements in science was his invention of the mass spectrometer. He was also a professor and later a good friend of Ernest Rutherford, who succeeded Thomson in the Cavendish chair.

In Thomson's 1913 paper, he reported that he'd been performing experiments similar to Ramsay's and Collie and Patterson's, but he wrote that his results favored a *different explanation* from transmutation. Thomson wrote that his method of analysis was more sensitive to the method used by the chemists.

Joseph John Thomson

Thomson reported that he, too, found traces of helium and neon, but he opted for the occlusion hypothesis. The only isotope of helium known to exist at the time was the dominant one, helium-4 (natural abundance 99.999863%). He also reported something far more significant, his discovery of a third mysterious gas:

> I may say that the primary object of my experiments was to investigate the origin and properties of a new gas of atomic weight 3, which I shall call X_3, which I discovered. ... This gas, as well as one with an atomic weight 20 (neon?), has appeared sporadically on the photographs taken in the course of the last two years; the discharge in the tube being the ordinary discharge produced by an induction coil

through a large bulb furnished with aluminum terminals, and containing gas at a very low pressure. (Thomson, 1913, 645, *Nature*)

Thomson also noticed that X_3 was produced most abundantly when he used platinum cathodes, part of the platinum metals group. Thomson knew that X_3 was very difficult to explain. "It is interesting to note," Thomson wrote, "that X_3 does not appear to occur to any appreciable extent in the atmosphere."

The significance of the discovery of this gas, with an atomic weight of 3, can be understood from today's perspective. There are two possibilities: It was either the extremely rare, stable helium-3 isotope, or it was tritium, the unstable hydrogen-3 isotope.

The existence of helium-3 was not even suggested for another 21 years, until Australian nuclear physicist Mark Oliphant proposed it in 1934. But the spectrum of helium-3 would have been qualitatively distinguished from that of hydrogen-3 (tritium).

Not until 1934 did the team of Rutherford, Mark Oliphant and Paul Harteck knowingly create and identify tritium. The half-life of this radioactive isotope of hydrogen, the time it takes to lose half of its radioactive power, is only 12 years. The tritium in the atmosphere before the era of atomic weapons existed only in trace amounts and came from rare cosmic radiation events. Tritium has since been produced in sizable quantities as a key component used in the nuclear weapons programs. Thomson did not realize the importance of his finding a gas that did not exist appreciably in the atmosphere. Without knowing it, Thomson produced the best evidence to confirm elemental transmutations with the cathode-ray method. As far as I know, it is the first recorded production of man-made tritium.

For his alternative explanation to Collie and Patterson's claim, Thomson decided that the helium-4 and neon had been embedded within the electrodes during the process of manufacture and that the gases were simply occluded within the electrodes until released by the cathode-rays. Patterson had melted the aluminum and found no occluded helium or neon, but Thomson did his own test. His results also contradicted his occlusion explanation.

Thomson put a piece of lead in a quartz tube and boiled it in a vacuum for three or four hours until three-quarters of it had vaporized. He examined the gases given off during the boiling process, and neither X_3 nor helium-4 could be detected. He took the remaining piece of lead and bombarded it with cathode-rays (energetic electrons). As expected from any successful science experiment, it showed direct cause and effect: he got appreciable amounts of X_3 and helium-4. Thomson didn't seem to recognize what he had done.

There is a close parallel to this failure to recognize experimental discoveries, and certainly many others presumably exist. As discussed in Chapter 3, Abel Niépce de Saint-Victor discovered radioactivity in 1858, but the discovery was not recognized by the scientific community until Antoine Henri Becquerel re-discovered it in 1896. A year earlier, in 1895, Roentgen had discovered X-rays.

Thomson made no suggestions about how the gases could possibly be embedded into the metals during manufacture. He also thought that cathode-rays had the capacity not only to penetrate metals but also to push out occluded gases from inside the bulk metals. He didn't see the inconsistency between this hypothesis and the fact that no gases were evolved from the metals even if they were vaporized. Here is where Thomson developed his ideas of occluded gases in his paper:

> I would like to direct attention to ... an everyday experience with discharge tubes — I mean the difficulty of getting these tubes free from hydrogen when the test is made... Though you may heat the glass of the tube to melting point, [you] may dry the gases by liquid air or cooled charcoal, and free the gases [that] you let into the tube as carefully as you will from hydrogen, you will still get the hydrogen lines... even when the bulb has been running several hours a day for nearly a year.
>
> The only exception is when oxygen is kept continuously running through the tube, and this, I think, is due, not to lack of liberation of hydrogen, but to the oxygen combining with the small quantity of hydrogen liberated, just as it combines with the mercury vapor and causes the

disappearance of the mercury lines. I think this production of hydrogen in the tube is quite analogous to the production of X_3, of helium-4, and of neon. (Thomson, 1913, 647, *Nature*)

Like other scientists of his day, he observed the difficulty of removing all traces of hydrogen from the surfaces of metals. Removing all hydrogen from what we now know are hydride-forming metals is extremely difficult.

By this time, scientists had known for at least eight years, thanks to the work of Clarence Skinner, that many metals easily absorb and retain hydrogen. They knew that fully removing hydrogen from metals was nearly impossible, once the metal had been exposed to a source of environmental hydrogen, such as air. (Skinner, 1905)

Thomson's mistake is revealed in his last sentence, where he assumed, without testing, that X_3, helium and neon behaved in the same way in metals as hydrogen did. Later research confirmed that helium is not soluble in and does not permeate intact, defect-free metals. (See Appendix B). Thomson did, however, fail to appreciate facts that were right in front of him: No helium was released even when he boiled the metal, and helium re-appeared when he again subjected the metal to cathode rays. (Thomson, 1913, 647, *Nature*)

Transmutation Despised

Despite the holes in Thomson's attempt to explain away the anomalous gases, Leonard Keene Hirshberg, the physician, scientist and writer I mentioned in Chapter 2, was sufficiently emboldened to go on the attack against the chemists' transmutation claims. Hirshberg began his 1913 article by referring to Ramsay's 1907 transmutation claim. Hirshberg lamented that Madame Curie had "disputed" Ramsay's results but that, sadly, she had no impact on discrediting Ramsay's claims. Hirshberg wrote that, even though Curie was recognized for her expertise in radium, Ramsay's "eminence as a research worker of undisputed renown overshadowed her."

Fortunately for science, other factors besides prestige sometimes serve to adjudicate acceptance or rejection of claims. Curie had, in fact, found nothing to dispute about Ramsay's claim; she merely failed to replicate the weakest part of his claims. But Hirshberg thought that Thomson had enough clout to pour cold water on the newer transmutation claims:

> Sir William Ramsay's position was so secure that, when [Ramsay] promulgated the startling discoveries, it sent a cold shiver up and down the spines of all but true-hearted believers. To overthrow the results of so eminent a savant must perforce require even, if possible, a greater investigator.
>
> That very situation has now come to pass. Sir J.J. Thomson, who is the discoverer of those tiny corpuscles called electrons, myriads of which are contained in every atom, who is also the Cavendish Professor of Physics at Cambridge University, England, and the leading investigator on electricity, magnetism, and light alive to-day, comes forward to deny Sir William Ramsay's conclusions unequivocally. These were based upon the presence of lines in the spectrum, a procedure known as spectrum analysis.
>
> Sir J.J. Thomson used an entirely different method which is much more sensitive than was Sir William's. ... The Cavendish professor [offered] an entirely different explanation for the presence of new elements as described by his London confrère. ...
>
> From Sir J.J. Thomson's detailed and painstaking experiments, the deduction is comprehensive and inevitable that the hopes of the alchemists are just as chimerical to-day as they were fifteen years ago. The elements supposed to have been transmuted in recent years by a little band of modern chemists are, after all, gases and elements actually present, yet hitherto unfound, hidden away deeply in the interstices of the metal from which they emanate. The bombarding rays from a [cathode-ray] tube merely liberate

the imprisoned elements, a power that even the greatest heat lacks. (Hirshberg, 1913)

Hirshberg was even cynical about the natural disintegration of radium into helium. "Moreover," Hirshberg wrote, "Sir J.J. Thomson has again given pause to the radium enthusiasts, for they, too, must rearrange their theories and their researches, and see, perhaps, whether radium itself really breaks up or merely expels the gases that are in it."

Burst of Helium

Physicist George Winchester (1875-1960) was next to try his hand at cathode-ray transmutation. Winchester grew up on a farm in Illinois and left school at 15. He finished his studies on his own, and after passing the state exam at 17, he began teaching in a one-room schoolhouse. A few years later, he became principal of the high school in Chillicothe, Illinois — the first time he entered a high school.

George Winchester

He went to the University of Chicago in 1899 to study physics and earned his Ph.D. in 1906. After a year at the University of Washington at Seattle, from 1907 to 1919, he was a professor of physics and head of the Physics Department at Washington and Jefferson College in Washington, Pennsylvania. During World War I, he was assigned as a scientific liaison to the Army Signal Corps and conducted experiments in the front-line trenches in France. In 1921, he returned to academia and was appointed chairman of the Rutgers University Physics Department.

His tenure at Washington and Jefferson College ended, according to the *New York Times*, "in a dispute over academic standards versus football prominence that gained national attention." The *Times* said he aroused the ire of the alumni association by flunking some leading athletes, and the alumni demanded his dismissal. Upton Sinclair, in his *The Goose Step — a Study of American Education* (1922), commented, 'Winchester had raised the money for the only class laboratory at the college, but he committed the offense of putting studies ahead of football, and for this, he was punished.'" (*New York Times*, 1960)

Winchester's working hypothesis, which he reported in 1914, was similar to the thesis of Thomson, and it had two components. The first was that the various metals he tested somehow held a hidden store of helium, neon and hydrogen. The second was that the electrons from the cathode-rays somehow could penetrate and force out the gases from the metals.

Winchester had experience with high-voltage cathode-ray tubes. Ten years earlier, he started performing experiments with the tubes, experimenting with ultraviolet light, and noticed the production of gases. A few years later, he ran more experiments to see whether he could remove all gases from the electrodes. Here's what he found:

> One thing noticeable in tubes of this kind is that whereas some yield only comparatively small amounts of helium, others are very rich in this gas. One tube, which at first gave only a small amount of helium in comparison with hydrogen, suddenly, after running for fifteen days, gave out an enormous amount of helium for a few days and then, just as suddenly, became normal again. ...
>
> The gases were measured and examined by the photographic plate; exactly the same gases were given off

from the two sets of electrodes in approximately equal amounts. Analysis, by means of the spectroscope, of the gases produced, shows the presence of hydrogen, helium and neon. (Winchester, 1914, 288-9)

He repeated the experiment again in 1913 and again detected hydrogen, helium and neon. He found that, if he ran the experiments for a month, the spectra of the helium and neon gases *disappeared.* He interpreted the cessation of helium and neon production as evidence against the transmutation theory and for the occlusion theory. (Winchester, 1914, 289)

Winchester assumed that conditions were constant in the tube during the month. He had no reason to consider at this early stage that hydrogen may have been a reactant in low-energy nuclear transmutations and thus was consumed during the month. As with Thomson, it was too early for Winchester to realize that hydrogen readily permeates hydride-forming metals but helium does not.

Winchester considered, though, that the production of helium and neon might have had something to do with their presence on the surface of the metals. He noticed a relationship "between the amount of gas evolved and the amount still remaining within the surface layer." Ninety years later, quite a few researchers found evidence that low-energy nuclear reactions took place on the surfaces of metals.

Nothing but Argon

Next in the line of helium and neon hunters came the work of Thomas Ralph Merton (1888-1969), an English physicist, inventor and art collector. His specialty was spectroscopy and diffraction gratings. When Merton published a paper in 1914, he was working in his private laboratory in his house. A few years later, he earned a D.Sc. from Oxford University and was appointed a lecturer in spectroscopy at King's College, London.

He found no helium or neon; however, he intermittently found argon gas in his experiments. He used aluminum electrodes treated with hydrogen "in order to wash any traces of other gases completely from the apparatus."

Even though he found "no trace of nitrogen," he did not propose that the argon was from a transmutation. Instead, he suggested that it could "be accounted for by an exceedingly small but continuous leak." Nitrogen constitutes 78 percent of air while argon constitutes less than 1 percent. Merton's explanation didn't make sense. Then again, transmutations without high-energy alpha radiation didn't make sense, either; he reported his results as negative. (Merton, 1914, 549)

A Most Useful Apparatus

Collie responded to Merton; he was not nearly as disappointed. Merton, presumably at Collie's request, transferred his apparatus to Collie, who performed experiments with it. Collie tried something new with Merton's apparatus. He bombarded finely powdered metallic uranium with cathode-rays. He found neon and helium but no argon.

It is unclear why Collie selected uranium as a target material. If he was looking to produce helium, uranium would be a poor choice for a starting material because it was a producer of helium nuclei, otherwise known as alpha particles. Helium, as a whole atom, is not a daughter product of the uranium decay network, but uranium does emit alpha particles during its decay which are helium nuclei. This fact would make it difficult to confidently ascribe his observed helium to transmutation rather than by a disintegration product from the uranium. It may be possible to calculate the amount of helium and neon he recorded during the experiments, typically lasting 90 minutes, and compare that with the expected amount of helium from the emitted alpha particles. I have not attempted such a calculation. The absence of argon was strong evidence for Collie that there had been no atmospheric contamination.

"If the neon and helium found were due to an air leak," Collie wrote, "it is difficult to account for the [absence] of the argon, which should have been present to the extent of one thousand times as much as the neon and helium found. The amount of argon present, however, was too little to be measured." (Collie, 1914, 554-6)

Searching for Hidden Gases

In November 1914, Collie, Patterson and Masson published their final paper on this series of experiments. The paper provided a comprehensive review of their work and, for the first time, provided diagrams of their experimental apparatus. In addition, for the first time, they mentioned that they ran many experiments that produced either no helium or no neon. These null results are significant because they mean that, as the researchers explained, they did not understand all of the key variables. Patterson also reported more blank (control) experiments. He set up an identical experiment but did not apply the electric discharge. The three authors explained a variety of the tests they performed. The two most convincing tests were these:

> In many of the experiments, the total volume of gas used was so small that even if it had been all atmospheric air, it could not have accounted for the quantities of the neon, and still less for those of the helium, which were obtained. ... The most obvious [test], which was exclusively relied upon in our earlier work, was the search for nitrogen. In air, the ratio of nitrogen to neon is about 80,000:1; hence if the neon detected in an experiment came from air, the nitrogen accompanying would be found in relatively overwhelming quantity. Trials in which very minute amounts of air were deliberately admitted and tested in the usual way verified this. (Collie, Patterson, and Masson, 1914, 40)

As did Thomson, these authors also vaporized an electrode to look for occluded gases:

> This has been further studied, however, by examining aluminum. It has been melted *in vacuo* and the evolved gases tested. They contained no neon or helium. To make perfectly certain, however, a fresh piece, about 3 inches long, of the same wire as was used for electrodes in successful

> experiments, was allowed completely to dissolve in potassium hydroxide solution in an apparatus whence all traces of air had first been removed. The hydrogen was passed directly over hot copper oxide; thence the residual gases could be washed directly into the large dead-space of a "non-transference" apparatus with the help of a little oxygen from a sealed-on permanganate tube. There was neither neon nor helium in the gas. (Collie, Patterson, and Masson, 1914, 44)

After the authors performed the tests and confirmed that the metals contained neither helium nor neon, the only other source they could imagine for pre-existing gases was the parts of the apparatus made from glass. And they melted samples of the glass, *in vacuo*, but found no helium or neon. They mentioned something curious when they discussed how they melted some of the glass to analyze for occluded gases. They wrote that one of the glass sources had a green color to it, and to them, this meant that its origin was "from Egypt and presumably at least 1500 years old; the other was an opaque yellow glass (Chinese, Keinlung period), about 150 years old. Neither specimen yielded neon or helium."

The authors described a replication attempt by Robert John Strutt (1875-1947) (the son of John William Strutt.) Strutt did not obtain any helium or neon in his experiment. Therefore, Strutt asserted, Collie and Patterson's results were the result of contamination, and, he wrote, they probably failed to safely transfer the contents of their tubes between the experimental and analytical apparatus.

Given the extensive tests they performed, Collie and his co-authors decided that only a minimal response to Strutt was required: "Of course, [we], from the very beginning of the work, were keenly alive to the fact that the slightest leak in the apparatus meant neon and helium in the end-product. Obviously, therefore, [we] took every precaution to prevent it."

The First Nuclear Transmutations

This full collection of experiments performed by Collie, Patterson, Masson, Merton, Thomson and Winchester demonstrated a broad, independent set of investigations that gave evidence for the first confirmed man-made nuclear transmutations. (Again, the term "nuclear" was not used yet.) Also, the transmutations were produced by low-energy processes, rather than by high-energy alpha particle bombardment. At the end of their paper, Collie, Patterson and Masson summarized their conclusion:

> We have endeavored to put the facts of the case as fully as possible, without reference to any preconceived theory. It is not our view that our experiments rigidly exclude all the possibilities which have been mentioned; but it is evident that the trend of the results is toward conclusions which, if they turn out to be true, would be of very obvious importance. (Collie, Patterson, and Masson, 1914, 45)

One more paper on the subject published in 1915. Alfred Charles Glyn Egerton (1886-1959) trained at University College, London, and briefly worked under William Ramsay. Egerton worked for the British Ministry of Munitions during World War I, then moved to the Clarendon Laboratory at Oxford, and later to the Imperial College of Science. Egerton wrote that Strutt and Thomson "failed to replicate" the helium and neon claims. Egerton then reported that he attempted his own replication and that he too failed to get any positive results.

Many factors can account for failures to replicate results. For example, Winchester pointed out an important difference between his experiments and some of Thomson's experiments. Winchester phrased his explanation of the production of helium and neon gases as being "eliminated" from the electrodes:

> Thomson describes an experiment in which he sparked iron electrodes in an atmosphere of 3 cm. of hydrogen for an

hour or so each day for four days; at the end of that time, not a trace of helium or neon could be found.

In my experiments, I have never been able to eliminate these gases so quickly from aluminum electrodes; using one aluminum electrode of 10 mg. weight as cathode with a platinum electrode of about 5 mg. weight as anode, helium has been discovered after several weeks of sparking almost constantly, [that is], in the neighborhood of eight hours a day. (Winchester, 1914, 289-290)

To my knowledge, these experiments have not been repeated since the 1920s. In the 1930s, the use of alpha particle bombardment, soon followed by various types of particle accelerators, allowed for much more repeatable, controllable experiments and better characterization of applied energies.

In contrast, low-energy transmutation phenomena were not theoretically understood at the time, and the experiments were difficult to repeat. Generally, low-energy transmutation experiments were assumed to be in error. For these reasons, the broad collection of work was omitted from all subsequent science books and largely forgotten. Such experiments could be repeated today with much better analytical techniques.

CHAPTER 14

Out of Gas

After Dramatic Headlines, Interest in Chemists' Transmutations Fizzles (1913-1927)

Now let's look at how the 1913 transmutation claims from high-voltage electric discharge experiments showed up in the newspapers. Ramsay, Collie and Patterson had announced the transmutation results at the Chemical Society meeting in London on Feb. 6, 1913. A reporter from the London-based *Morning Post* was the first to break the news on Friday morning, Feb. 7. Hours later, the news appeared on page 1, column 1 in the *New York Times.*

The news marked the first set of confirmed man-made nuclear transmutations, and Ramsay, Collie and Patterson diplomatically shared the credit for it. Ramsay presented his paper first and reminded his fellow chemists that, just a few years earlier, he had obtained lithium from copper. The *Morning Post* reporter wrote that the audience was mildly incredulous to the point of laughter at Ramsay. But they took Collie seriously, and he was applauded several times during his report.

The reporter wrote that, although the scientists confidently measured transmuted elements, they had no clear explanation for the responsible process or processes. As far as the scientific method goes, this was not a criticism; the method does not require a viable theory as a condition of acceptance of an experimental claim. The *Post* writer reported that the authors suggested, among the various possible explanations, that hydrogen may have been the source. (Nearly a century later, this was proved correct.)

> # FIND ELEMENTS ARE TRANSMUTED
>
> Three British Scientists Have Apparently Done This or Evolved Matter from Energy.
>
> ## NEON AND HELIUM PRODUCED
>
> From Substances in Which They Were Not Previously Known to Exist.
>
> ## ELEMENTS SURELY CHANGED

Headline from page 1, column 1, Feb. 7, 1913, New York Times *article*

The reporter recognized that the cathode-ray experiments signaled a new line of research that was distinct from all prior radioactivity research. "At any rate, one thing seemed certain," the reporter wrote. "The elements could be changed, and they could be changed in a way very different from the way that radium was changed. In its case, the process could neither be hastened nor retarded. But the present phenomenon was artificial, and, further, the process was occurring at the other end of the system of the atoms, producing elements of low atomic weight. The old idea of the transmutation of elements had to be altered. We were coming to know more of subatomic matter, and it had to be realized that [quoting Alfred Tennyson] – "the old order changeth; yielding place to new, and God fulfills Himself in many ways, lest one good custom should corrupt the world."

Arthur Smithells

After the scientists read their papers at the meeting, Arthur Smithells (1860-1939), a British chemist and professor at the University of Leeds, led the discussion, according to the *Morning Post* writer:

> Professor Smithells opened the discussion by saying that he was somewhat breathless at the papers they had heard. It required a great deal of courage for scientific workers to bring forward such results, and [the audience] must admire it. Their courage, he thought, was justified, for their

experimental record was such as to justify what [from other researchers] would have been extremely rash. ... The obvious criticism was that in the [experiments], enormous weight had necessarily been laid on the spectroscopic evidence, and his limited experience in this connection had taught him caution, but he felt sure that authors were too experienced to fall into such pitfalls.

Sir William Ramsay expressed his great gratification at other researchers having taken up the investigation. With radium, there had been no chance of repetition, but the present experiments on transmutation could be reproduced by anyone with a coil and battery. (Morning Post, 1913, 155)

The point about radium is worth explaining: Radium, particularly the highly refined and most potent material used by the Curies, Rutherford, and Ramsay, was not available to everyone who wanted it. It was a rare, precious material. The cathode-ray experiments presented no such barrier, and this explains the many replications that took place during a 24-month period.

Nothing to Stand On

Oliver Joseph Lodge (1851-1940), a prominent British physicist, was not quite convinced. Lodge was a professor of physics and mathematics at University College, Liverpool and later became the first principal of Birmingham University.

"Though it is very enticing to suppose that atoms can be manufactured out of ether or that complex atoms can be built up out of simple ones, a great deal of proof will be required before it can be accepted," Lodge said.

Although it is a fairly common practice, Lodge's use of the word "proof" was a subjective response, not a scientific response. Proof is unattainable in science. There could never be proof, let alone a great deal of it. A German technology developer who has a background in science summarized this concept eloquently in a letter to me.

Real science does not try to prove things. That is not possible. It tries to disprove things, trying to learn from errors, then having a chance to propose new hypotheses. If something is not disproved, it is accepted as current common agreement to be true, as long as nobody comes along and finds an experiment to prove it as false. New scientific hypotheses and theories always allow for new experiments to disprove them.

It's not fair to criticize either Thomson or Lodge for failing to understand the significance of Thomson's finding of X_3 because they did not know what tritium was at the time. The foundation of Lodge's critique rested on Thomson's speculation that the helium and neon gases that he observed had been occluded in the metals, presumably because the alternative explanation would have been so extraordinary. Thomson speculated that the rare gases had been included within the metals during the manufacturing process, though he did not support this with any facts. He speculated that the cathode-rays had a special power to expel the gases from the metals. Although he knew that metals retained hydrogen, he did not confirm that neon or helium behaved the same way. His only test, melting a piece of lead, contradicted his occlusion theory. Thomson, as well as Lodge, had nothing to stand on.

Productive and Protective

A brilliant editorial appeared in the *New York Times* on Feb. 11, 1913, that eloquently captured the interplay between the scientists that placed themselves on one side or the other of the transmutation dilemma:

> In considering the difficulties which [many] well-qualified scientists have found in accepting as even probable, the transmutation of elements and the changing of energy into matter, [a person] must give weight to relative eminence of the proponents and critics of these provisional hypotheses and also take into account what may be called the

temperaments of the two.

It is not to be charged to Sir William Ramsay as a fault that he is known to be [a person] who makes use of his imagination in interpreting his observations. Imagination has often been of enormous service when brought to bear on the misty borderland between knowledge and ignorance, and to it, science is almost as much indebted as is art. The old [1800s] theory of atoms and molecules and the supposedly immutability of elements — these, too, are products of the scientific imagination. In other words, they are merely assumptions of long-proved convenience, and it was only their age that gave them the appearance of "laws" or "truths."

Sir William Ramsay and his associates knew as well as do their critics of the possibility that the gases and solids of mysterious origin that appeared in the course of their experiments may have been lurking in the molecular or atomic interstices of their apparatus. They also know, for they explicitly admitted it, that perhaps this was the source of what they found, but they had a feeling or impression that they had done something strange and new, and the intuitive conclusions of such men, being based on a vast amount of research and information, all severely exact, cannot be lightly dismissed.

The great leaders in science are audacious and cautious at the same time, and the one quality is as productive as the other is protective. The one thing for which they have no patience is authority, and they are as ready to assail an axiom as a guess. (*New York Times*, 1913, Imagination)

The author's perspective on controversial science is as valid today as it was then. However, scientists, science journalists, and science enthusiasts are far more deferential to science authority now than they were a century ago, when so many rapid advances were made in the newly discovered field of radioactivity.

Interest Fizzles

The interest in transmutations in cathode-ray tubes fell rapidly in the 1920s. In 1913, in his series of annual progress reports, Soddy considered and conditionally accepted the Collie and Patterson claims, but the following year, Soddy put more emphasis on the "negative" findings of Thomson and Merton. (Trenn, 1975, 354-6, 383-5) Most historians did the same. With the exception of Stewart, they didn't perform their own analysis, instead relying on and recording the history based on the perspective of Thomson.

World War I began in 1914, and most pure science research came to a halt; many young men went to war, and other scientists were assigned to military tasks or applied research. There is no evidence of any cathode-ray experiments performed until after the war ended in late 1918. A few scientists in the early 1920s attempted but failed to reproduce the production of gases described in the previous chapter. (Bureau of Standards, 1922, 12) They suggested that the cause of the 1912-14 claims was contamination from atmosphere by passage of helium through glass walls, even though the earlier authors explicitly tested for it and used double-walled chambers.

In 1925, R. W. Riding and E.C.C. Baly published a paper in which they tried to test the influence of the different types of discharges that the earlier researchers had used. They also tried to see whether they could find a repeatable method to create helium or neon. I found that paper confusing and was unable to draw any clear conclusions from it.

However, in October 1926, the pair published a letter in *Nature*, reporting results of additional experiments that produced small amounts of helium and neon. They had confirmed the results of Collie, Patterson and Masson. (Baly and Riding, 1926)

A few weeks later, Robert W. Lawson responded that Baly and Riding did not sufficiently consider the permeability of helium and neon through glass. However, Lawson's speculation seems irrelevant because Baly and Riding observed successful as well as null results by varying experimental parameters; but all tests, as far as I can tell, were performed in glass.

One scientist who suggested the contamination idea was A. Lo Surdo, but, according to Joseph William Mellor, the editor of many voluminous compendiums of chemistry research, another scientist, E. Cardoso, even though he failed in his replication attempt, showed that Lo Surdo's explanation was unsatisfactory. (Mellor, 1927, 948)

And Now, Gold

The scientific research between 1912 and 1914 inspired a revival of the interest in the ancient traditions of alchemy. In late 1912, Herbert Stanley Redgrove (1887-1943) founded the Alchemical Society, and he published the short-lived *Journal of the Alchemical Society* from 1913 to 1915.

On February 16, 1913, the *New York Times* reported that it had been contacted by Rudolph Melville Hunter of Philadelphia, who claimed that he had been producing massive quantities of gold for 10 years. The *Times* asked Collie for his opinion, and he dispensed with the claim deftly. He said that people who claim the ability to make gold commercially would not need to seek media attention or investors. (*New York Times*, 1913, Notes)

A month later, the *Sunday Times* (London) published an article with a greatly exaggerated headline: "COMING DOOM OF GOLD: TRANSMUTATION OF METALS." Here's an excerpt:

> If the constituents of wood could be changed into copper, it is clear that the market for that metal would be seriously affected and would become unstable. If iron or other commonly occurring substances could be changed into the so-called noble metals, we should no longer be able to employ gold as the standard of value, and the whole monetary system of the world would collapse. (*Sunday Times*, 1913)

However, only helium and neon had been created by transmutation. No scientist had confirmed any transmutation of metals — yet.

CHAPTER 15

Rutherford Shatters the Atom

The First Man-Made Disintegration of a Stable Atom Generates an Abundance of Alchemical Myths (1919)

The electron-induced reactions reported between 1912 and 1914 revealed the first confirmed man-made nuclear transmutations. Although the *New York Times* published a misleading headline suggesting that inexpensive base metals had been transmuted into gold, only noble gases had been produced. The next event on the transmutation timeline was about alpha-induced reactions.

A Science Fiction

According to most history books, Internet sources, and even a comic book produced on behalf of the Atomic Energy Commission to promote atomic energy, Ernest Rutherford performed the first man-made nuclear transmutation. As the story goes, in 1919, he bombarded alpha particles and transmuted nitrogen gas into oxygen gas. That story is fiction.

Looking back, he may very well have done such a transmutation, but his experiments showed no data to support the idea, he published no such claim and it was not his intention. He did, however, bombard alpha particles into nitrogen gas, although his hypothesis, his data, and his claim had nothing to do with the creation of oxygen. Rutherford's fictional transmutation is one of the greatest myths of nuclear science history. It was not Rutherford but his graduate student Patrick Blackett, in 1925, who earned the right to the nitrogen-to-oxygen claim.

Busting Atoms

Here is what Rutherford actually did. His intention was to see whether he could bust apart the atom into some of its constituent parts. There is no evidence that he intended to transmute elements. In fact, his aversion to transmutation is evident in his comment to Soddy, "They'll have our heads off as alchemists." Rutherford hypothesized that emitted alpha particles had the necessary power to break up stable atoms. This wasn't a revolutionary idea, but nobody had demonstrated it convincingly. Ramsay claimed to have accomplished this in 1907, but no one had confirmed his experiment independently.

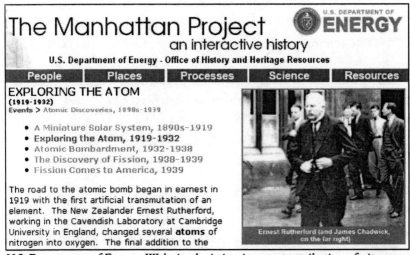

U.S. Department of Energy Web site depicting incorrect attribution of nitrogen-to-oxygen transmutation, retrieved June 23, 2014

Rutherford's friend and mentor J.J. Thomson spoke about the idea in a June 10, 1914, lecture. "The bombardment by alpha rays," Thomson said, "seems to be the most promising means of producing atomic transformation." (Trenn, 1974, 64) Alfred Egerton also mentioned the idea in a 1915 paper. "It is possible the alpha particle does possess sufficient energy to disintegrate some of the atoms it encounters (see Cameron and Ramsay, 1907)," Egerton wrote. (Egerton, 1915, 188)

A science historian at the time, Alfred Walter Stewart, a professor of chemistry at Queens University of Belfast, explained why Rutherford's hypothesis was easy to imagine. Stewart had published a series of reviews called *Recent Advances in Physical and Inorganic Chemistry*.

> Ramsay was the first to see that the energy of the alpha particle offered our best chance in this field; and Rutherford's recent researches have proved the accuracy of Ramsay's insight. The alpha particle is a submicroscopic projectile with a diameter estimated at 8×10^{-13} cm and a speed of about 10,000 miles per second in its free path. Mass for mass, its energy is four hundred million times as great as that of a rifle bullet. It is evident that, when such a body is suddenly brought to a standstill by collision with matter, the energy-change will far exceed anything which can be produced artificially on the same scale. (Stewart, 1926, 149)

Comic strip frame from an Atomic Energy Commission-sponsored publication (General Electric Co., 1948)

According to science historian Thaddeus Trenn, Rutherford first observed his discovery of such an event on November 9, 1917.

Rutherford wrote a letter to Niels Bohr explaining what he was trying to do: "I am also trying to break up the atom by this method." (Trenn, 1981, 69)

What Happened Next

If Rutherford did not observe the transmutation of nitrogen into oxygen between 1917 and 1919, then what exactly did he do? He used a product of the decay of radium to shoot alpha particles into nitrogen gas and saw that protons were ejected. In other experiments, he used oxygen instead of nitrogen. He assumed, incorrectly, that the alpha particle knocked a proton off the nitrogen or oxygen nucleus on impact. Chapter 22 will explain what actually happened. Somehow, he also assumed, correctly in this case, that the emitted proton did not come from the fractured alpha particle. For years leading up to 1920, the proton was referred to as a hydrogen atom, or H-particle. The neutron had not yet been discovered, either. Here is Stewart's description:

> When alpha particles from a radioactive source are allowed to pass through an atmosphere of hydrogen, it is found that scintillations can be observed on a fluorescent screen which is placed beyond the normal extreme range of the alpha particles. This implies that under these conditions something is striking the screen; and since the screen and the radioactive source are separated by a distance which no alpha particle can span, it is evident that this "something" cannot be an alpha particle.
>
> The explanation of the phenomenon appears to be as follows:
>
> To take an extreme case, let us imagine that an alpha particle meets a hydrogen atom "end-on" and that a collision occurs. Since the hydrogen atom is four times lighter than the alpha particle, it is evident that it will recoil after the collision at a speed much greater than that possessed by the alpha particle; and being endowed with this greater velocity, it will be able to

travel through a longer distance than the original alpha particle would have done before coming to rest at the end of its normal range. It is therefore assumed that the scintillations on the screen are produced by these recoiling hydrogen atoms, which are termed H-particles.

Rutherford has examined the phenomenon in the case of gases other than hydrogen, and in the case of nitrogen, he obtained an anomalous effect. When a stream of alpha particles was passed through chemically pure nitrogen, it was found that the range of the recoiling atoms, instead of being small — as might've been expected when the relative masses of the alpha particle in the nitrogen atom are compared — was comparable to that which would have been produced if hydrogen had been present instead of nitrogen. In fact, the penetrating power of these particles was found to be even greater than that of the H-particles from hydrogen.

If these particles should eventually be proved to be H-particles, there seems to be no escape from the idea that the nitrogen atom has been shattered by the impact of the alpha particle upon it; and that hydrogen is one of the disintegration products. ... Should the Rutherford particles eventually prove to be hydrogen atoms, it will establish the correctness of Ramsay's view (which was rejected at the time by more than one expert in radioactive problems), that the energy of radioactive bombardment might suffice to disintegrate atoms which normally are stable. (Stewart, 1920, 196-7)

I've provided a list of Rutherford's publications related to this discovery in Appendix D. He first presented his disintegration of nitrogen to hydrogen in a lecture at the Royal Institution on June 6.

In a detailed and meticulous set of four papers, he published his findings in the June 1919 *Philosophical Magazine*. Those papers are dense and technical; however, the following year he summarized his work when he presented his Bakerian Lecture on June 3, 1920, in London at the Royal Society. He summarized his achievement in one paragraph:

> While it has long been known that helium is a product of the spontaneous transformation of some of the radioactive elements, the possibility of disintegrating the structure of stable atoms by artificial methods has been a matter of uncertainty. This is the first time that evidence has been obtained that hydrogen is one of the components of the nitrogen nucleus.

He also made clear what he did not do. He knew that he had not figured out what happened to the target nucleus after it ejected the proton. He knew that figuring out what happened to the remaining target was the natural next step of inquiry. "It is natural," Rutherford wrote, "to inquire as to the nature of the residual atoms after the disintegration of oxygen and nitrogen, supposing that they survive the collision and sink into a new stage of temporary or permanent equilibrium." (Rutherford, 1920)

He speculated about a mechanism, which, as he also wrote, was limited by insufficient data available at the time and, as it turns out, was wrong:

> The expulsion of an H atom carrying one charge from nitrogen should lower the mass by 1 and the nuclear charge by 1. The residual nucleus should thus have a nuclear charge 6 and mass 13, and should be an isotope of carbon. If a negative electron [*] is released at the same time, the residual atom becomes an isotope of nitrogen.
>
> The expulsion of a mass-3 carrying two charges from nitrogen, probably quite independent of the release of the H atom, lowers the nuclear charge by 2 and the mass by 3. The residual atom should thus be an isotope of boron of nuclear charge 5 and mass 11. If an electron escapes as well, there remains an isotope of carbon of mass 11. The expulsion of a mass-3 from oxygen gives rise to a mass-13 of nuclear charge 6, which should be an isotope of carbon. In case of the loss of an electron as well, there remains an isotope of nitrogen of mass

13. The data at present available are quite insufficient to distinguish between these alternatives. (Rutherford, 1920)

[* Before the discovery of the neutron, scientists thought the nucleus was composed of protons and electrons.]

A Nobel Myth

In his 1919 series of papers, Rutherford reported that, by using alpha particles, he was able to cause the disintegration of a stable atom. Sometimes, he ran experiments using nitrogen as a target and, other times, using oxygen *as a target*. As far as he knew, the disintegration emitted a proton. That's it. He never speculated that the *residual atom* from the impact might be an oxygen nucleus. As shown in the previous paragraph, he hypothesized that the residual atoms would be either boron or carbon.

Rutherford had no intention of making a claim of any transmutation, let alone a transmutation of nitrogen to oxygen. He intended merely to be the first person to bust apart a nucleus of a stable atom, and in this he succeeded. Beyond that, however, he was very clear that his available data were "quite insufficient" to determine the nature of the residual nucleus. He assigned that task to his graduate student Patrick Blackett, who worked for the next four years, got the data and reported it in 1925. Blackett made the transmutation claim, provided the correct analysis, and published the supporting data in 1925. I will review Blackett's work in Chapter 23.

The mythology of this incorrect history extends even to the Nobel Foundation itself. As of June 2014, here's what the organization said on its Web site:

> In 1919, during his last year at Manchester, [Rutherford] discovered that the nuclei of certain light elements, such as nitrogen, could be "disintegrated" by the impact of energetic alpha particles coming from some radioactive source, and that during this process fast protons were emitted. Blackett later proved, with the cloud chamber,

that the nitrogen in this process was actually transformed into an oxygen isotope, so that Rutherford was the first to deliberately transmute one element into another. (Rutherford, 1920, 385)

The statement "Blackett later proved" implies that Rutherford made the initial transmutation claim, which he didn't. The statement "Rutherford was the first to deliberately transmute one element into another" is false. Furthermore, it is inconsistent with his character. Of all scientists, Ernest Rutherford would have been the last person to try to transmute one element into another. Changing elements was not Rutherford's goal or interest; he was interested in the nature of particles and atomic composition.

Helium Transmutation Myth

A second myth, which persisted for only a few years in the early 1920s, is that Rutherford had transmuted hydrogen into helium. I do not know how Rutherford came to this erroneous view. After Rutherford performed his experiments with targets of nitrogen and oxygen, he performed experiments with many other target elements and found that they, too, emitted protons on impact from alpha bombardment.

Rutherford reported these findings in a paper in *Nature* on May 6, 1922. In this paper, he also led people to think that he had produced helium from these experiments. "These results," Rutherford wrote, "show clearly that the nuclei of heavy atoms contain both positively charged helium nuclei and negative electrons, and lead to the general view that the complex nuclei of all atoms are built up of hydrogen and helium nuclei and electrons." (Rutherford, 1922, 584, *Nature*)

But Rutherford had not transmuted any hydrogen into helium. His experimental results *did not clearly show* that helium was a constituent of all atoms; this was conjecture. Nevertheless, the editors of *Nature* in May 1922 labeled Rutherford's disintegrations "contemporary alchemy." "We think it no exaggeration," *Nature* wrote, "to say that these experiments are some of the most fundamental which have ever been made. It is not often a scientific discovery excites interest outside the narrow circle of the laboratory or the scientific lecture."

Rutherford certainly proved that a wide spectrum of elements were composed of hydrogen, what we would now call protons, and this served as a major foundation from which to continue the developing understanding of the atom. But transmutation to helium? No. Transmutation of nitrogen to oxygen? No, that was Blackett's accomplishment. We resume the story in Chapter 22 with Blackett's 1925 transmutation.

CHAPTER 16

Golden Rumors

Rumors Spread That Rutherford Turned Lead into Gold and Germans Manufactured Artificial Gold (1920-1922)

This chapter examines two sets of intriguing news stories. The first stories published in late 1919 and were follow-up stories to Rutherford's 1919 disintegration discovery, though fact and error in the stories were commingled. The second set of stories was published in January 1922. The stories described rumors that German scientists were learning to transmute inexpensive materials into synthetic gold.

There are not enough facts about the German claims to allow evaluation of the claims themselves. But this chapter gives readers a sense of the times and the opportunity to view the transmutation conversation through the eyes of someone who might have been reading the newspapers back then.

"Transmutation" of Nitrogen to Hydrogen

In Sept. 19, 1919, an article appeared in *English Mechanic and World of Science*, a weekly newspaper that was published from 1865 to 1926. The article discussed the September 1919 annual meeting of the British Association for the Advancement of Science.

The author of the article mentioned Rutherford's June 1919 claim of producing hydrogen from nitrogen and wrote that his discovery seemed to be of great importance. Neither Rutherford nor anyone else at the time distinguished between hydrogen (the element) and the proton (a subatomic particle.) It was too early for that distinction. That's why some people thought that Rutherford had transmuted one element into another, nitrogen into hydrogen. (*English Mechanic*, 1919)

Masthead of English Mechanic and World of Science

January 1920 News

On Jan. 15, 1920, a U.S. magazine, *Youth's Companion*, ran a follow-up story based on the *English Mechanic* story. The 120-word article, "Learning About Nitrogen," said, "Recent experiments with the alpha ray have led to the discovery, says the *English Mechanic and World of Science*, that nitrogen, which for a century and a half has been regarded as an element, may not be an element at all, but a compound of hydrogen and helium."

As we know today, nearly everything in that last sentence was incorrect. Nitrogen is an element, not a compound. Nitrogen is an element composed of protons, neutrons and electrons. Rutherford's conclusion that he had knocked off "hydrogen" from nitrogen was the only part of the sentence that was fairly accurate. It was, of course, a proton that was emitted from the reaction. But he actually didn't knock the proton off the nitrogen atom. The exact mechanism was revealed in 1925, and we'll get to that in Chapter 22. Rutherford's suggestion that helium was, in some way, a constituent of nitrogen was a guess, and wrong.

The *Youth's Companion's* news wasn't recent; Rutherford had reported the experiments half a year earlier. Nevertheless, the *Youth's Companion* story got picked up by a wire service and was syndicated, appearing in perhaps thousands of papers nationally. (*Youth's Companion*, 1920)

April 1920 News

A few months later, on April 2, 1920, the *Canton Times* published a story titled "The Beginnings of Transmutation?" It wasn't quite as triumphant as the other stories:

> To read some of the press comments on the news of a recent achievement by Sir Ernest Rutherford, one would think that he had discovered a way of changing lead into gold or dust into foodstuffs. ... If Sir Ernest has succeeded in obtaining hydrogen from nitrogen, this proves not that nitrogen has been "transmuted" into hydrogen in the old alchemic sense, but that nitrogen is a compound of which hydrogen is an element. (*Canton Times*, 1920)

In hindsight, this was a lot closer. A proton is certainly a constituent of nitrogen. Rutherford never claimed that he transmuted nitrogen into a different element, though some historians have fictionalized Rutherford's history as though he did.

May 1920 News

On May 30, 1920, the *Boston Globe* ran a story titled "TURNING LEAD INTO GOLD: Oxford Professor Thinks Alchemists of Old Had Right Idea." The most interesting part of that news story is the following sentence: "Sir Ernest Rutherford is devoting himself ... to experiments designed to convert lead into gold." (*Boston Globe*, 1920)

Of course, Rutherford was the last person in the world who would have run experiments designed to convert lead into gold. So what was Rutherford actually doing after his 1919 discovery? The *Times of London* got it mostly right.

August 1920 News

On Aug. 25, 1920, the *Times of London* reported on the Aug. 24 meeting of the British Association for the Advancement of Science. The *Times* quoted W.A. Herdman, a professor of oceanography at the University of Liverpool, who gave the presidential address.

Herdman first talked about the synthesis of helium from hydrogen that took place in the stars. He discussed the potential heat liberated from that reaction and speculated that, if the reactions were to occur on earth, they would "more than suffice for our [energy] demands."

"Sir Ernest Rutherford had recently been breaking down the atoms of oxygen and nitrogen," Herdman said. "If combinations requiring the addition of energy could occur in the stars, combinations which liberated energy ought not to be impossible." (*Times of London*, 1920)

What *Was* He Doing?

After Rutherford published his papers in 1919 reporting the disintegration of nitrogen and oxygen into hydrogen, he and his colleagues did similar experiments with many other target elements.

Their interest was the disintegration of elements: characterizing the particles emitted from the targets. They didn't try to characterize nor did they intend to characterize the identity of the residual nuclei and to look for newly transmuted elements.

January 1922 News

Now we move forward to 1922. Rutherford's 1919 discovery was still getting attention. It wasn't just the newspapers that were describing his disintegration work as transmutation. Even some of his peers were quick to label Rutherford's achievement a man-made transmutation.

On Jan. 8, 1922, the *New York Times* published "WAY TO TRANSMUTE ELEMENTS IS FOUND: Dream of Scientist for 1,000 Years Achieved by Dr. Rutherford." The *Times* ran the story as though Rutherford demonstrated that he had transmuted elements:

> The transmutation of elements, the dream of both charlatans and scientists for nearly 1,000 years, has actually been accomplished by the recent work of Sir Ernest Rutherford, and his results are generally accepted by scientists and physicists, according to Dr. James Kendall, associate professor of chemistry at Columbia, who said, on the other hand, that there was not the slightest reason to believe that the Germans had accomplished their reported feat of making synthetic gold. (*New York Times*, 1922, Way to Transmute)

Rutherford never made any claim of transmutation although it is easy to see how in 1922 it could have seemed that way. We'll get to the rumors of the synthetic gold in just a minute.

The *Times* wrote that Rutherford's discovery had possible far-reaching implications, and it quoted professor O. W. Richardson, who had recently given a presidential address to a Mathematics and Physics section meeting of the British Association for the Advancement of Science. Richardson foresaw a new age of nuclear weapons and energy:

> The artificial transmutation of the chemical elements is thus an established fact. The natural transmutation has, of course, been familiar for some years to students of radioactivity. The Philosopher's Stone, one of the alleged chimeras of the medieval alchemists, is thus within our reach. But this is only part of the story. It appears that in some cases the kinetic energy of the ejected fragments is greater than that of the bombarding particles. This means that these bombardments are able to release the energy which is stored up in the nuclei of atoms.
>
> Now we know from the amount of heat liberated in radioactive disintegration that the amount of energy stored in the nuclei is of a higher order of magnitude, some millions of times greater, in fact, than that generated by any chemical reaction such as the combustion of coal. ... If these effects can be sufficiently intensified, there appear to be two

possibilities. Either they will prove uncontrollable, which would presumably spell the end of all things, or they will not. If they can be both intensified and controlled, then we shall have at our disposal an almost unlimited supply of power which will entirely transcend anything hitherto known. It is too early yet to say whether the necessary conditions are capable of being realized in practice, but I see no elements in the problem which would justify us in denying the possibility of this. It may be that we are at the beginning of a new age, which will be referred to as the age of subatomic power. We cannot say; time alone will tell. (*New York Times*, 1922, Way to Transmute)

Rutherford failed to see the potential of atomic energy.

Germans Make Artificial Gold!

Well, maybe not. But that's what the *New York Tribune* reported on December 31, 1921. (*New York Tribune*, 1921)

The Jan. 8, 1922, *New York Times* story, which had heavily promoted Rutherford for achieving the transmutation dream, briefly mentioned the German gold claim. It quoted James Kendall, an associate professor of chemistry at Columbia, who called it nonsense.

A few weeks later, the German gold news was depicted more dramatically. On Jan. 21, 1922, the *Olean Evening Herald*, in Olean, New York, ran the headline "Synthetic Gold: Roger Babson Shows Havoc German Chemists Might Play in Financial World" in its version of the story. (*Olean Evening Herald*, 1922)

A writer from journal *Mining and Scientific Press* explained one of the reasons why the news was believed. "Such a story was given currency by two persons generally reputed to have good sense — Mr. Irving Fisher, a professor at Yale, and Mr. Roger W. Babson, a press writer on financial affairs," the journal wrote. (*Mining and Scientific Press*, 1922)

Here's how Babson introduced his story:

> Private advices have been coming to me for some time that German chemists are diligently working to discover some method for making synthetic gold. I do not know that the German government is directly behind such experiments, but indications point in that direction. The German government would of course be justified in spending a tremendous sum of money to discover a process of making synthetic gold.

Headline from Jan. 21, 1922, Olean Evening Herald

Babson was certain that an international conference would be called for immediately to assess the potential impact, and he mentioned in his article a Yale University professor, Irving Fisher, as a financial expert. As it turns out, Fisher was Babson's source. The contents of the article were as dramatic, and as devoid of facts, as its headline.

Fast Government Reassurance

Did the United States government take this seriously? You bet. On Jan. 18, 1922, the United States Geological Survey, perhaps after doing some research, issued a statement. Here's the core of it:

Modern chemistry has shown that at least some of the supposedly elemental substances of the chemist, what he calls elements, are in fact compounds. In all ordinary chemical processes, these compounds behave like elements, but it is nevertheless possible by special chemical operations to show that they are divisible into more simple substances. This discovery has revived, to some extent, popular belief in alchemy, and there have been of late many suggestions in the press that gold may be made artificially and become so abundant as to destroy completely such utility as it may have as a measure of value and a basis for currency. (U.S. Geological Survey, 1922)

Fast Retraction

Seven days later, on Jan. 25, Fisher did a complete turnaround. He told the *Times* reporter that he had been misled by the German man, whom he did not identify in any way.

GOLD MAKER A FAKE, PROF. FISHER FINDS

German Alchemist Turns Out to Have a Prison Record, Yale Economist Reports.

SOME GOLD FROM THE SEA

German Government Made Serious Attempts to Extract Gold From Water During the War.

Headline from Jan. 25, 1922, New York Times *article*

The German alchemist turned out (suddenly) to have a prison record and was a fake, according to an interview Fisher gave to the *Times:*

> Berlin, Jan. 24 – Prof. Irving Fisher of Yale University, who recently caused a sensation by stating that a certain German scientist had apparently discovered a method of producing synthetic gold, today abandoned his treasure hunt. I had a long talk with the professor this afternoon when he told me that up to this forenoon his search appeared promising, but on account of information he received an hour or two before the interview took place, he was convinced it was in vain. (*New York Times*, 1922, Gold Maker)

In the shadow of Rutherford's valid disintegration discovery, the fear of a major disruption to the U.S. financial system was, at least to some people, conceivable. In the next chapter, we go back to the hydrogen-to-helium transmutation idea, and in Chapter 20, the gold story resumes.

CHAPTER 17

Hotter Than the Sun

Danger: Exploding Wires!

In 1922, at the University of Chicago, two chemists accomplished nuclear transmutation with a relatively simple table-top apparatus. They used an experimental method known as the exploding electrical conductor, also known as an exploding wire. This chapter provides a basic introduction to the phenomenon. Chapter 18 discusses the details of the Chicago chemists' work in more detail.

History of the Exploding Electrical Conductor Phenomenon

Hundreds of scientific papers have been written about the exploding electrical conductor phenomenon, and at least three bibliographies on the phenomenon are in the public domain. The field began with exploding wires and expanded to include the use of foils and films.

The most comprehensive of these is the 157-page bibliography and abstract compilation, a fourth edition produced in 1967, by William G. Chace and Eleanor M. Watson at the former U.S. Armed Services Technical Information Agency. (Chace and Watson, 1967)

The second bibliography is a 19-page index from 1966 compiled by James R. McGrath at the U.S. Naval Research Laboratory. This bibliography includes a useful review of the most notable research on the phenomenon. McGrath wrote that the first, most-extensive spectrographic analysis on the exploding electrical conductor phenomenon was performed around 1920 by John August Anderson, an astronomer working at the Mount Wilson Observatory, in California. Exploding wires have certain fascinating luminosity characteristics.

"[Anderson] pointed out," McGrath wrote, "that the absorption spectrum [of exploding wires] is more complete than any other laboratory source, that the brightness is greater than the sun, and that the continuous background extends into the extreme ultraviolet region." (McGrath, 1966)

The third bibliography, of 35 pages, was produced by William H. Richardson in 1958 for Sandia Corporation under contract to the federal government. (Richardson, 1958)

Chace explained in a 1963 article in *New Scientist* that there was no universal agreement on exactly what was happening to cause wires to behave in such a manner:

> An overloaded fuse wire usually melts quietly. If, however, a wire is deliberately subjected to a sudden and very high current, it can explode violently. This phenomenon is still not fully understood, 200 years after its discovery, but it is usefully applied in photography, metalworking and research. (Chace, 1963)

Steve P. Hansen, a former semi-conductor engineer who now specializes in vacuum technology and is the creator of the belljar.net Web site, wrote a brief article explaining the basics of exploding electrical conductors in 1993, and it is presented with his permission here.

Basic Concept

Here is how Hansen explained the basic concept of the exploding-electrical conductor phenomenon:

> The first thought that might come to mind about the topic of exploding wires could be what happens when a fuse

blows: a brief flash of light, perhaps accompanied by a slight pop. The two phenomena are fundamentally different. Whereas the fuse wire melts with some associated electrical arcing as the wire forms into hot droplets of metal, in the exploding wire process, electrical energy is dumped into the wire at such a high speed that the wire detonates into a plasma.

In physics, a plasma is known as a fourth state of matter, and it resembles a hot gas, which consists of positively charged ions and negatively charged electrons. When the exploding wire bursts into a plasma, it's accompanied by a sonic shock wave.

The standard method for exploding a wire is to feed energy to it through the discharge of a high-voltage capacitor. The use of an electrical discharge to heat or vaporize metal wire, foil or films was first described in the 18th century by Martin van Marum (1750-1837), a multifaceted Dutch natural philosopher. Marum performed research in various disciplines, including botany, geology and medicine as well as electricity, and was able to melt 70 feet of thin wire using the discharge from a battery of Leyden jars (an early form of an electrical capacitor). But this was not yet an example of the exploding-wire phenomenon. Throughout the following years, a progression of experiments took place: first with warm wires, then evaporating wires and finally, as soon as researchers had the ability to produce larger current pulses with capacitors, exploding wires.

The ever-ingenious Benjamin Franklin investigated one of the early aspects of the research. Franklin coated cardboard silhouettes with gold and silver leaf and used a Leyden battery to vaporize the foil. When he placed the silhouette close to a piece of paper, he reproduced the image of the silhouette on the paper. Franklin's experiment was still not a true exploding wire.

An everyday example of Franklin's experiment happens

when an incandescent light bulb blows; a small amount of tungsten vaporizes and leaves a deposit on the glass envelope. Likewise, when a glass fuse blows, vaporized metal is left on the inside of the glass. According to Frank Früngel, author of a 1965 book on the topic, Edward Nairne (1726-1806), an English instrument maker and a friend of Franklin's, was, in 1774, the first person to report the true exploding-wire phenomenon. (Source: Früngel, 1965)

For most of the 19th and 20th centuries, the exploding-wire phenomenon was a mild curiosity for scientists, but it became the subject of intense research in the 1950s. These explosions, too brief for the wire to fall apart before complete vaporization, were used to generate very high temperatures (tens of thousands of degrees), initiate shock waves, produce dense plasmas, act as light sources for high-speed photography, and even form sheet metal into complex shapes.

Currently, exploding wires are being used in energy research. "Wires" made of frozen deuterium are fed into vessels where very-high-powered discharges are used to attempt nuclear reactions. As fuel, the frozen fibers are more efficient than deuterium gas for producing dense plasmas. A well-known example is the Sandia National Laboratories' Z-pinch experiment. (Hansen, 1993)

The Explosion Process

For readers who would like a more-technical explanation, here is how Hansen explained the explosion process:

> In the case of the fuse, the current rise is usually fairly slow (the better part of a millisecond). First, the wire softens from a solid to a thick liquid. Then, the molten metal forms into a chain of droplets as a result of the action of surface tension. When these droplets separate from each other, the fuse breaks.

Usually, there is arcing between the droplets before the circuit is finally broken.

Alternatively, the exploding-wire phenomenon depends on a very fast current rise, on the order of a few microseconds. Rather than breaking at a few points, the entire wire changes state simultaneously from a solid to a plasma. Frank Früngel described the process as comprising six steps:

1. Heating of the wire by the heavy current passing through it.

2. Formation of a liquid column of metal replacing the wire.

3. Development of instability in the wire, which forms unduloids (figures of rotation) and causes the appearance of striations in the metal vapor. (This is from the mechanical and magnetic forces created by the discharge and heating.)

4. On the disruption of the arcs formed between the unduloids, a dark interval results, during which no current flows through the wire; the voltage remains constant.

5. A sudden flash of light is observed, the spectrum of which is continuous and independent of the material used. Thereafter, emission and absorption lines are observed, depending on the materials used.

6. One or more sonic shock fronts are created during the same interval. (Source: Früngel, 1965, 379-90)

Below is a series of six consecutive high-speed X-ray flash photographs by Frank Früngel, showing an exploding iron alloy wire. The exposure time for each frame was 0.3 microseconds, and the film in the camera was moving at 10,000 feet per second. The photograph clearly shows the initial distortion of the wire, then the centers of the explosion. (Hansen, 1993)

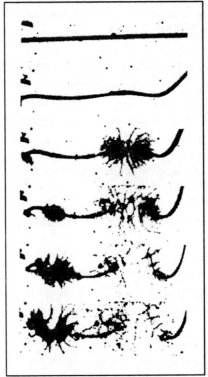

*Flash X-ray photographs of an exploding constantan wire
(Source: Früngel, 1965, 389)*

It Goes Boom

As many videos available on the Internet show, an exploding-wire experiment is relatively easy to set up and takes only a few minutes to demonstrate. The Massachusetts Institute of Technology Department of Physics Technical Services Group has published such a video; it is available at the group's Web site, http://techtv.mit.edu/videos/635, as well as on YouTube at https://www.youtube.com/watch?v=-3IbAerYj8I.

Below is a single frame from the video during the middle of the explosion:

Frame from video capture of exploding-wire experiment at MIT

A True Explosion

Chace's *New Scientist* article provided more evidence that the phenomenon was a true explosion rather than a melting wire. He showed a photograph of a wire that was to be exploded, surrounded by a large copper tube. The force of the exploding wire was strong enough to rip apart the tube.

Different diameter wires, he wrote, produced different explosive sound profiles. A relatively large (0.020 inch diameter) wire produces a deep and cannon-like sound. A thinner (0.001-0.005 inch) wire produces a sharp sound that is painful to the ear. The resulting cloud was not smoke, Chace wrote, but metal oxide:

Exploding wires result in a cloud of metal vapor at extremely high temperatures and intense electromagnetic fields — conditions which result in a highly ionized state. Here is a plasma but one unlike most plasmas; its material density is high. Hence, exploding wires are very important as a source of dense plasma. It is this plasma property which makes the exploding wire a potential means of rocket propulsion.

The very high current densities attainable have already led to experiments in which Ohm's law broke down in metals. Further uses await only new techniques for working in nano- instead of micro-seconds and the ingenuity of a future generation of physicists.

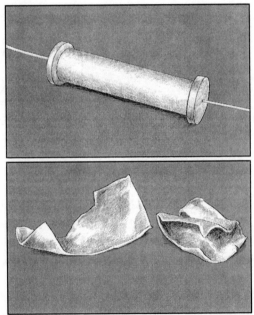

Sketch by Arturo Aguirre of a copper pipe before and after destruction by an exploding wire, based on a photograph in Chace's New Scientist *article.*

Exploding wires can have enough kinetic energy, and have been used, to shatter concrete foundations. (Urutskoev, 2002)

CHAPTER 18

Exploding Wires in Chicago

Wendt and Irion Enter Forbidden Land of Transmutation, Explode Tungsten and Make Helium (1922)

Whereas Rutherford, contrary to popular myth, had not transmuted hydrogen into helium, two chemists, in 1922, at the University of Chicago, did accomplish the 1,000 year-old dream of alchemy.

Paradoxically, whereas the famous physicist received credit for a transmutation he neither observed nor claimed, the less-famous chemists received marginal recognition for their transmutation observation and claim. Rutherford, in fact, was no impartial party; he went on the attack to discredit the chemists' claim even before their paper published.

In the last chapter, we covered the basics of the exploding-wire phenomenon. Before we get into the details of the 1922 Chicago exploding-wire experiment and transmutation, let's recall what had happened in the previous few years. In 1912, several chemists had reported the transmutations of hydrogen to helium and neon, triggered by low-energy electrons within cathode-ray tubes.

Sir Joseph John Thomson, then holding the prestigious Cavendish chair at the University of Cambridge, tried to discredit the chemists' transmutation research with his occlusion hypothesis, and he largely succeeded.

His hypothesis was that the newly found helium and neon gases in the cathode-ray experiments were not the result of transmutations but that the gases had somehow been implanted in the metal cathodes before the experiment and somehow coaxed out of the metal during the experiment.

But Thomson did not test the hypothesis; he simply speculated. Nevertheless, for most scientists then, the occlusion hypothesis was easier to believe than the idea of an alchemical phenomenon.

In November 1914, Collie, Patterson and Masson explicitly tested Thomson's occlusion hypothesis and disproved it. On the other hand, Thomson's tritium production, even though he didn't realize it, was the best confirmatory evidence in support of elemental transmutation.

In 1919, Rutherford successfully shattered a nitrogen atom with the high energy of the alpha particle and reported the emission of a proton.

Forbidden Land of Transmutation

On March 11, 1922, chemists once again entered the forbidden terrain of transmutation. Two scientists at the University of Chicago made a surprising announcement. at an American Chemical Society meeting. The next day, the news went nationwide, probably worldwide. It was a first: Chemists at the Kent Chemical Laboratory of the University of Chicago said, according to news reports, that they had transmuted, at will, a metal, tungsten, into another element, helium.

The senior chemist was Gerald L. Wendt (1891-1973), 31 at the time; the junior chemist was Clarence E. Irion (1896-1976), 26 at the time. They didn't use the word transmutation, but they reported their results as the artificial decomposition of tungsten into helium. (Wendt and Irion, 1922)

Stolen Spotlight

Rutherford knew only that he had busted apart atoms. He knew that he lacked information about the change to the residual elements and therefore was not in a position, even if he wanted to, to claim a transmutation.

As mentioned in Chapter 15, Rutherford did not, contrary to myth, report a transmutation of nitrogen into oxygen. His hypothesis was that hydrogen was a constituent particle of the nitrogen atom, and indeed, his experiment emitted a hydrogen particle. In later years, scientists understood that the hydrogen particle was a proton.

> **ENGLISH PROFESSOR**
>
> Solves Age-old Riddle of Changing Metals Into Gold.
>
> Paris, Dec. 8—Sir Ernest Rutherford, since 1907 Langwordthy professor and director of physical laboratories at the University of Manchester, England, has solved the riddle of transmutations of metals, the secret sought by the ancient alchemists, according to the Matin.
>
> Sir Ernest Rutherford is one of the best known physicians in the world. He has devoted much attention in recent years to radio activity.
>
> Transmutation of metals, one into another, especially the conversion of baser metals into gold, has long been sought. Sir Ernest is believed to have arrived at his conclusions through experiments with radium, to which he has devoted considerable attention in recent years.

1919 Associated Press news story published in the Madison Daily Herald *perpetuated the erroneous idea that Rutherford had transmuted metal. He did no such thing. He caused each of the gases nitrogen and oxygen to emit another gas, hydrogen.*

Fundamentally, though, Rutherford was correct in that he observed the emission of a proton. He also observed similar "hydrogen particle" emissions when he shot alpha particles at oxygen nuclei targets. After 1919, he performed similar experiments with a variety of other target elements and found that these bombardments also emitted hydrogen particles. Rutherford developed the erroneous belief that helium, as well as "hydrogen," was a constituent subatomic particle of all atoms. He published his final articles from that series of experiments in May 1922. Before he did, however, Wendt and Irion stole the headlines from him on March 11.

Imagine the situation from Rutherford's perspective; he could not have been pleased. Here was Rutherford, holding the prestigious Cavendish chair, recognized worldwide for solving the ancient riddle of the alchemical transmutation of metals, knowing full well that he had done no such thing. Now, two relatively unknown American *chemists* come along and report that they changed the metal tungsten into another element, helium, at will.

A Few Words About Irion

Irion, the junior researcher, had worked on the experiments while he was an undergraduate student. He completed his bachelor's degree at the University of Chicago in electrical engineering in 1921. The following year, he went to Iowa State College and earned a master's degree in metallurgy and chemistry. He earned his Ph.D. in organic and physical chemistry from Ohio State University in 1935 and worked as an industrial chemist for a variety of companies throughout his 55-year career.

World-Famous Wendt

Wendt's life in science, on the other hand, was quite prominent. His claim of decomposition of tungsten into helium was far from the highlight of his career, and he certainly did not try to overstate his and Irion's claim. In the summary of their paper, they wrote, "We wish to emphasize that this report is preliminary, and that nothing is proved beyond the importance of the problem and the promise of this method."

Casual research reveals little about the man. But the news archives offer plenty. Although Wendt has been mostly forgotten, he was once considered a top science authority in the United States, if not the world.

Wendt was born in Davenport, Iowa, in 1891. He attended Harvard University and earned his bachelor's degree in chemistry, his master's degree in chemistry and physics and his Ph.D. in physical chemistry. His research at Harvard included investigating the behavior of alpha-rays on hydrogen. From 1912 to 1916, he was an Austin teaching fellow at Harvard. After Harvard, he went to work for the Bureau of Mines in 1916 as a junior chemist in radioactivity and worked on radium production. The bureau sent him to Paris to study radioactivity with Marie Curie. According to one of Wendt's books, *You and the Atom*, he became the first U.S. expert on radium.

Gerald L. Wendt in 1944

Wendt served in World War I as a captain in the Chemical Warfare Service, where he worked on the development of toxic gases and the design of gas masks. During his career, he was an instructor of chemistry at Rice Institute in Houston, an instructor in quantitative analysis and radioactivity at the University of Chicago and, later, an associate professor there. He was appointed dean of the School of Chemistry and Physics at Pennsylvania State College. He also worked for industry and was awarded several patents.

Despite its potential significance, the transmutation work Wendt did with Irion was not a continuing passion for him. Sometime between 1921 and 1922, Wendt experienced a serious health problems and had to halt his research. In 1922, Irion left for Iowa State College. In their paper, Wendt and Irion wrote that Irion planned to continue the research at Iowa, but I haven't found any evidence that either of them continued the research.

Wendt's greatest joy was inspiring and educating the public about the wonders and benefits of science. For many decades, in addition to his activities as a professor, and at times working for industry, Wendt enjoyed touring the country and giving lectures to the public and science teachers. He wrote hundreds of news articles on science for the general public, was science editor for *Time* magazine, a consultant for *Life* and *Fortune* magazines, and the editor of several science magazines. As a science consultant for CBS-TV, he produced short movies that explained everyday applications of science.

A 1958 news story in the *Thomasville (Georgia) Times-Enterprise* said that Wendt's science columns appeared in more than 5,000 newspapers in 71 countries and that he was often called on to communicate science to the masses and to journalists:

> [Wendt] is generally considered to be the best-informed scientific writer in the United States and one of the best speakers in the country appearing on the platform. A distinguished American scientist before entering the writing and motion picture fields, Wendt is considered America's foremost interpreter of scientific progress.

Dr. Wendt ranks high in the field of atomic energy, as is indicated by the fact that his latest book, *You and the Atom*, published in January 1956 already has been translated into 14 foreign languages. In recognition of such, it was Wendt who, at the "Atoms for Peace" conference in Geneva, analyzed the atomic discussions for the benefit of the 900 newspaperman present. (*Thomasville Times-Enterprise*, 1958)

In his service to professional journals, he was the associate editor of the *Journal of the Radiological Society of North America* for some time and the editor-in-chief of one of the most important chemical industry publications, *Chemical Abstracts*, for 12 years. He was also the editorial director of *Science Illustrated* and editor of *The Humanist*.

Wendt held the post of director of the American Institute of the City of New York, a civic organization founded in 1828 and dedicated to the promotion of American industry, agriculture, and science. Perhaps his most interesting role in the public communication of science was as the director of science and education for the 1939-40 New York World's Fair. He spent most of the time from 1940 until his retirement in 1967 working in science publishing.

In 1951, when he was 60, Wendt began working for the United Nations Educational, Scientific and Cultural Organization (UNESCO) in Paris. In 1956, he returned to the United States and set up the UNESCO Publications Center in New York, which later became the National Agency for International Publications Inc.

Wendt wrote eight books: *Matter and Energy: An Introduction by Way of Chemistry and Physics to the Material Basis of Modern Civilization, Volume 1*, with Oscar Franklin Smith (1930); *Science for the World of Tomorrow* (1939); *The Atomic Age Opens* (1945); *Atomic Energy and the Hydrogen Bomb* (1950); *Nuclear Energy and Its Uses in Peace* (1955); *You and the Atom* (1956); and *The Prospects of Nuclear Power and Technology* (1957). He was the editor of *Chemistry: The Sciences, a Survey Course for Colleges* (1942).

Now let's go back to the day he found himself in the spotlight.

50,000 DEGREES HEAT DECOMPOSES METAL

Chicago Chemists Claim Transmutation of Tungsten to Helium Gas.

USE ARTIFICIAL LIGHTING

Create Most Intense Heat Known in Universe by Explosion of 100,000 Volts of Electricity.

CHICAGO, March 11.—Transmutation of metals, sought throughout the ages, has finally been accomplished, it was announced in a paper read today at a meeting of the Middle Western sections of the American Chemical Society at Northwestern University.

Headline from New York Times *news story on March 12, 1922*

Astounding Transmutation News

When Wendt and Irion first reported their results at the March 1922 American Chemical Society meeting, they deliberately avoided the word transmutation, but it made no difference. The next day, the newspapers depicted them as alchemists. "Transmutation of metals," the *New York Times* wrote, "sought throughout the ages, has finally been accomplished."

Technically, it wasn't a metal-to-metal transmutation like lead to gold. But it was reported as a transmutation of a metal to a gas. Unlike the 1912-14 experiments, which used high-voltage electric fields in cathode-ray tubes for hours or days continuously, the Wendt-Irion experiment took the form of an exploding thin wire, subjected to a single, split-second burst of high power.

Transmutation, Sought Throughout the Ages

Wendt didn't view his and Irion's work as transmutation. He called it decomposition. In his mind, he and Irion had done more or less what Rutherford had done in 1919, by splitting nitrogen and causing it to emit "hydrogen," only he and Irion started with tungsten instead of nitrogen, and they got helium instead of hydrogen. Here is how Wendt described his interpretation of the process in "Decomposing the Atom," an article he wrote for *The Nation*.

"Transmutation, 'sought through the ages,' as the daily press has it," Wendt wrote, "is too large a word to apply to our present subatomic powers. Disintegration is certain; decomposition is just becoming recognized; but integration and synthesis are nearly as far off as ever." (Wendt, *Nation*, 1922, 564,)

When Wendt wrote his article for *The Nation*, he did not accept as confirmed the 1912-14 transmutation work by Collie, Patterson and others. "Only one method is well-established," Wendt wrote, "that of Sir Ernest Rutherford. Another [method], which has been wrongly heralded as transmutation, has been applied in [my] laboratory with such promising results that confirmation by other investigators has become important."

Deafening Explosion

An August 1922 article in *Popular Science Monthly* gave a basic description of Wendt and Irion's experimental method:

> The method consists in charging a massive electrical condenser with the current of 30,000 volts pressure, and then rapidly discharging this large quantity of energy under high pressure through an extremely fine metallic wire. A

violent explosion results as the wire is vaporized. During the first 1/300,000th of a second, the light emitted was 200 times as bright as sunlight, according to determinations made with a rotating mirror. The pressure momentarily developed was between 50 and 100 atmospheres, or on the order of 750 to 1,500 pounds per square inch. The noise of the explosion was so loud that the ears of observers had to be protected. When the wire was exploded in a glass vessel, the latter was shattered into tiny fragments. When the explosion was conducted in a glass tube, under water, the wire completely disappeared, and the tube could be found only as an unrecognizable powder.

Another [exploding-wire] experiment [by other researchers] demonstrated how brief a time was involved in the explosion. The fine wire was wrapped in thin tissue paper, and the condenser discharged through it. The paper was torn to shreds, but no signs of scorching were discovered. (*Popular Science Monthly*, 1922)

Artist's depiction of an exploding-wire experiment in the August 1922 Popular Science Monthly

Inspiration From the Stars

Wendt explained in his article in *The Nation* that he wanted to go further than Rutherford. "Rutherford worked on a minute scale," Wendt wrote, "using individual alpha rays with a skill that, even for his unique genius, is phenomenal. ... Unquestionably, [his work] is atomic decomposition, though on an excessively minute scale and with the application of relatively tremendous energy."

Wendt got his inspiration from the stars. More precisely, he looked at the method developed by John August Anderson, an astronomer working at the Mount Wilson Observatory in California. I will allow Wendt to explain, in his own words from his article, his hypothesis, Anderson's method, and the potential significance of his and Irion's experiment:

> Chemists are limited to much cruder methods [than that used by Rutherford]. In our own work, we took a hint from the astronomers, who have found that the composition of the stars varies with their temperature. Judged by their spectra, they seem to contain the same elements as the earth, but it is significant that on the hotter stars many of the heavier and more complex elements are missing.
>
> The very hottest stars, with temperatures approaching 20,000° Centigrade, show predominantly the spectrum of helium, while the progressively cooler classes of stars reveal an increasing list of the heavier elements. The obvious suggestion is that, at extreme temperatures, the atomic collisions are so violent as to shatter the larger nuclei or to prevent their formation from the smaller units of hydrogen and helium. This is far from certain, however, because the effect of high temperature may be only complete ionization, complete separation of all the external electrons, since it is these electrons which produce the spectra, and without them we have no astronomical means of identifying the elements.

This effect of extreme temperature became capable of laboratory tests through the ingenuity of Dr. J.A. Anderson of the Mt. Wilson Solar Observatory of the Carnegie Institution. He devised a method for producing temperatures above 20,000° Centigrade, or 35,000° Fahrenheit, which consists of the discharge of a large quantity of electricity under high pressure through a very fine metallic wire. A massive electrical condenser is charged by a transformer to a voltage well above 30,000 volts, preferably 100,000. By suitable control, this energy is then instantaneously sent into a short wire about 0.001 inch in diameter. The heat generated vaporizes the wire and heats the vapor to a degree much hotter than any previously reached in the laboratory and apparently hotter than prevails on any star.

A brilliant flash, 100 times brighter than direct sunlight, is produced, lasting only 1/300,000th of a second. Dr. [Anderson] has used this method for the study of high-temperature spectra. We harnessed the explosion in such a way as to collect the gases formed. Tungsten wires were used, chiefly because the high atomic weight of that metal renders its decomposition probable if the hypothesis is correct.

The wires weighed about half a milligram, slightly over 1/100,000th of an ounce, and the gas produced in each explosion occupied a volume of about 1 cubic centimeter. *This represents the conversion of nearly half the material of the wire into helium, which is the chief product, according to the spectroscopic analysis of the gas.* [emphasis added]

This, then, if confirmed by later work, is atomic decomposition on a much larger scale than Rutherford's, though still small. It is significant, not in revising our conceptions in any way, for that had been done by Rutherford in 1920, but in opening the way for large-scale investigation. Further study should show what other elements were formed besides helium and, when a quantitative technique is attained, should give us fairly precise information as to the exact structure of the atomic nucleus.

And, perhaps even more important, it should determine the energy conditions within the nucleus and the availability of atomic energy for man's use. It is conceivable that it will lead to a synthesis of the smaller atoms into larger ones, and when that becomes possible, the manufacture of any metal from hydrogen and helium and from other metals will be within range — the transmutation that fascinated the mind of medieval baron and alchemist.

Energy, however, rather than gold, is the vital need of modern civilization, and it is energy which is the prime factor in the subatomic world. (Wendt, *The Nation*, 1922, 654)

A Lot of Gas

In their paper, published in September 1922, Wendt and Irion reported a series of 21 experiments, each producing a quantity of gas. They identified helium as the predominant element in the gas, and on average, the experiments produced 1.01 cc of gas, primarily helium, with a maximum of 3.62 cc in the first run:

> Abundant gas was present in the bulbs after explosion. Visual spectroscopic examination of it uniformly disclosed the faint presence of the strongest green line of mercury, probably from back-diffusion from the pumps. The only other line uniformly present and positively identified was the strong yellow line of helium. This was always brilliant when sufficient exciting current was used, though it gradually weakened after the current was passed for some time. This is characteristic of helium, which, under such circumstances, is easily absorbed by the electrodes.

Helium does not permeate intact, defect-free metals, so perhaps there was another explanation, such as catalysis, for the gradual disappearance of the helium after it was produced. Wendt and Irion knew that the inside of their chamber was not contaminated by air because they saw no hydrogen or neon spectral lines in the gas:

> Both hydrogen and neon were absent. The appearance of helium and the absence of hydrogen are interesting for two reasons. In the first place, it seems to dispose of the objection that the helium arose from gas remaining in the wire, for in that case, hydrogen should also have been visible, for it was probably originally present in the wire in much larger quantity than was helium. (Wendt and Irion, 1922)

They also ran experiments surrounded by carbon dioxide to further exclude the possibility of helium leakage from the atmosphere.

Thrust Into the Spotlight

Wendt and Irion, despite their effort to avoid the transmutation tag, failed. The Associated Press, which wrote and syndicated the story nationally the day after the American Chemical Society meeting, didn't care about Wendt and Irion's descriptive preference. In turn, Wendt and Irion didn't care for the Associated Press' characterization of their claim, as they wrote in their journal article.

"We regret the exaggerated earlier published account of this work," Wendt and Irion wrote, "which was given wide publicity through the Associated Press upon its oral presentation at an intersectional meeting of the middle-western Sections of the American Chemical Society at Northwestern University, Evanston, Illinois, on March 11, 1922."

The New York Times promoted the transmutation angle. The *Times* quoted Paul N. Leech, a member of the Chicago section of the American Chemical Society, who offered his perspective on the significance of the work:

It means that the alchemists who tried to turn the baser metals into gold were right on one point – that the nature of metals could be changed. But, of course, it has nothing to do with the assertions of scalawags that the baser metals can be transmuted into synthetic gold. It does, however, actually blast the theory that the atoms of elements, supposed to be absolutely indestructible, cannot be broken up by men. It opens up a vast new field to science and may result in many far-reaching and important scientific developments. We cannot yet foresee what these developments may be.

Up until 1895, it was believed that no decomposition of elements was possible. At that time, however, it was discovered that radium, which is one of the about 90 known elements, naturally decomposes into lead. Nature, however, performs that change, and until Dr. Wendt and Mr. Irion completed their experiments, man had never been able to produce a similar result. (*New York Times*, 1922, 50,000 Degrees)

Alchemy was back. However, Leech and the *Times* had overlooked the possible 1907 man-made transmutation by Ramsay, the man-made transmutations by half a dozen scientists from 1912 to 1914, and the 1919 man-made disintegration by Rutherford.

Two Atomic Giants

Journalists referred to the previously unknown Wendt in the same context as the now world-famous Rutherford. An August 1922 feature article in *Popular Science Monthly* mentioned both "atomic explorers" but featured Wendt's work prominently. The magazine began with Wendt's futuristic speculations that a bucket of fuel could power oceangoing ships and that a tiny amount of radium could light entire cities for dozens of years. Wendt also imagined that someday the sun would provide a direct source of solar energy, according to a Sept. 13, 1928, *Mason City Globe Gazette* article:

> The world of tomorrow will be a simple synthetic world — with lumber made from waste stalks; gems manufactured in a laboratory; silks, leathers, glass coming from cellulose; sugar, foods from the chemists' retort; fat, acids from petroleum; power, light and heat from the sun. Such is the prediction of Gerald L. Wendt, dean of Pennsylvania State College and a foremost authority on scientific research.

The news that "went around the world," as *Popular Science Monthly* described it, placed Wendt and Irion's discovery — "the transmutation of elements" — on par with, if not of greater significance than, Rutherford's discovery. Here are a few excerpts from the magazine:

> The present excitement over atomic decomposition was stirred up, primarily, by two different series of experiments — those undertaken by Sir Ernest Rutherford, the famous English scientist, and the recent phenomenal exploit of Wendt and Irion.

Atom Explorers Open Up New Worlds

THE most amazing activity of scientists today is the exploration of the atom. They are opening up a vast world of new knowledge that intimately concerns every other interest you may have, be it radio, electricity, astronomy, medicine, or what not. But do you understand the methods by which atom explorers are probing these minute solar systems, which boast of relatively staggering distances all comprised within almost infinitesimal space? Do you realize the industrial possibilities of their work? The above account of recent spectacular experiments in knocking atoms to pieces will illuminate for many readers a new and extremely fascinating realm of science that no well-informed man today can afford to ignore. — THE EDITOR.

Sidebar from August 1922 Popular Science Monthly *article*

> Can energy be used? In wonderfully delicate experiments, ... Rutherford may be said to have achieved atomic decomposition, but on an almost incredibly minute scale. Just how long it will be possible on a large scale to break up the atoms composing a given material, so that the

energy thus released can be turned to practical commercial use, is a problem yet to be solved.

The still recent but already famous experiment of Wendt and Irion was undertaken, however, in the hope of pioneering ... ultimate commercial utility, following a path different from Rutherford's. Their scheme was to decompose atoms by applying heat, rather than by bombarding them with the tiny projectiles given off by radium, as Rutherford had done. (*Popular Science Monthly*, 1922, 31)

Unlike Rutherford, Wendt was an avid enthusiast for the practical applications of science, an optimist about nuclear energy, as well as a great popularizer of science to the masses. Rutherford, on the other hand, was a career academic: well-known for his aversion to transmutation and for his pessimism about the prospects of atomic energy. With the exception of a short period during World War I when he worked on applied research for submarine detection, Rutherford spent his life in the academic world. *Popular Science Monthly* captured Wendt's contrasting perspective and vision in this quote, at the end of the article:

Atomic decomposition on a much larger scale than ever before attempted seems thus to have been attained. It is conceivable that ultimately it will lead to the synthesis of the smaller atoms into larger ones, and when that becomes possible, the manufacture of any metal from hydrogen and helium and from other metals will be within range — the transmutation that fascinated the mind of medieval baron and alchemist. (*Popular Science Monthly*, 1922, 32)

Wendt recognized that any source of energy could be used for destructive as well as constructive purposes. In a book he wrote in 1950, *Atomic Energy and the Hydrogen Bomb*, just months after President Harry S. Truman announced his decision to support its development, Wendt spoke about the bomb's potential destructive power.

Nevertheless, he did not attempt to stifle scientific and technological progress, as is clear from his article in *The Nation*:

> If its subatomic energy were available, a pound of radium would easily propel the largest liner across the Atlantic and leave nearly a pound of lead in place of cinders. It is, of course, not now available, for radium takes its own predestined time for this transformation, and it is reckoned [that it takes] thousands of years. But it is the great promise of atomic decomposition that [the] means will be found to liberate this energy from common elements at will. When that happens, the future coal supply will need worry no one, and coal strikes will at last be at an end. It is afar off, but a new industrial era can be pictured which makes the coal age seem medieval indeed. And beside that prediction, the prospect of ever making gold from dross fades into insignificance. (*Nation*, 1922, 564)

Rutherford Jumps the Gun

The Wendt-Irion experiment offered promise, hope, and world-changing potential. Their discovery suggested far more significance than Rutherford's. Rutherford wasted no time. Within a month, long before Wendt and Irion published their paper, Rutherford submitted identical letters to the editors of *Nature* and *Science*, based on what he heard about the experiment from the news media. Rutherford did not attempt a replication of the experiment. He merely offered theoretical arguments. (Rutherford, 1922, *Nature, Science*)

> **SCIENTIFIC EVENTS**
> DISINTEGRATION OF ELEMENTS[1]
> I HAVE been asked to say a few words about a telegram in the *Times* of March 14 giving an account of a paper communicated to the American Chemical Society at Chicago by Dr. G. Wendt and Mr. C. E. Iron.

First paragraph of Rutherford's letter critiquing the Wendt and Irion claim

By doing this, Rutherford put himself in a vulnerable position because he was responding only to news reports rather than a published journal article.

On May 26, 1922, Wendt responded to Rutherford's letter with his own letter to *Science*. Wendt was neither subtle nor timid with the great Rutherford:

> Sir Ernest Rutherford, in the statement copied from *Nature* in the April 21 issue of *Science*, was in the very difficult position of being "asked to say a few words" in comment on a brief cablegram to the *London Times*, which was itself based on an exaggerated Associated Press dispatch to American newspapers concerning the preliminary and oral but as yet unpublished report of Mr. Clarence E. Irion and myself on the apparent decomposition of tungsten at extremely high temperatures. [Rutherford] mentions the need of a complete report before intelligent comment is possible, but proceeds to make three points which are properly conservative and entirely correct but, as will be seen from the complete paper upon its publication in the *Journal of the American Chemical Society*, are all irrelevant. (Wendt, 1922, *Science*)

Rutherford had three criticisms, and Wendt revealed the deficiencies in each of them.

Occluded Helium

"During the last ten years," Rutherford wrote, "many experiments have been recorded in which small traces of helium have been liberated in vacuum tubes in intense electric discharges, and it has been generally assumed that this helium has been in some way occluded in the bombarded material."

Wendt responded that the existence of a decade of ambiguous results was not a scientific argument that offered direct critique against his and Irion's claim, nor was the matter of ambiguity a valid scientific critique of that previous work. Wendt wrote:

True, [there is] a list of no less than 37 papers, most of them published in the years 1912 to 1915, engaged in this inconclusive argument. In spite of the application of the best experimental skill, no agreement was reached, and Rutherford's conclusion [of occlusion in metals] is the general one. Yet there are some of the final experiments, particularly those of Collie, which challenge that conclusion, and the problem is still one of the most attractive and important of recent times. (Wendt, 1922, *Science*)

Wendt was obviously aware that, in their final set of experiments, Collie, Patterson and Masson had specifically tested Thomson's occlusion hypothesis and disproved it.

Theoretical Argument

"The disintegration of a heavy atom into lighter atoms, e.g., into atoms of helium," Rutherford wrote, "would be accompanied by a large evolution of energy. Indeed, it is to be anticipated that the additional heating effect due to this liberated energy would be a much more definite and more delicate test of disintegration of heavy atoms into helium than the spectroscope."

Here is Wendt's response to Rutherford's second point.

"This is a rare example," Wendt wrote, "of the preference for theory over fact, though saved by the use of the word 'test' instead of 'proof,' and the chemist will be slow to accept it. Lack of the theoretical energy does not explain away the formation of a cubic centimeter of permanent gas [helium] from half a milligram of tungsten wire."

Part of Rutherford's critique was an attempt to cast doubt on the helium because Wendt and Irion had made no heat measurement. Such data might have provided support for existing theory. However, Wendt had exposed the deficiency of Rutherford's critique: The scientific method dictates that experiments have priority over theory.

Power Comparison

Rutherford's third point was that normal use of X-ray tubes did not produce helium; therefore, he insinuated, the helium observed by

Wendt and Irion was an error.

"In [X-ray] tubes," Rutherford wrote, "an intense stream of electrons of energy about 100,000 volts is constantly employed to bombard a tungsten target for long intervals, but no evolution of helium has so far been observed."

Wendt explained that Rutherford's third argument was irrelevant because Rutherford was trying to make an invalid power comparison.

"Finally," Wendt wrote, "Sir Ernest points out that no helium has been observed in X-ray tubes operating at 100,000 volts, where electron impacts are even more violent than in our experiments. But the quantity of energy impressed on the target is here minute, the tube current being measured in milliamperes or less, whereas it is the essence of our method to introduce as much as a Coulomb of electricity into the wire within 1/300,000th of a second, or many millions of times as much in terms of power."

I have found no evidence that Rutherford responded in any way, either in a letter or in a formal comment, to the journal. To my knowledge, no scientist has ever examined the Wendt and Irion paper and found any error of experimental protocol or analysis or any crucial, unstated assumption. Other scientists, however, made varying degrees of effort to replicate the Wendt and Irion claim, but they all failed. Before we get to those reports, let's take a look closer look at the competition between Wendt and Rutherford.

Atomic Race Among Men

Radioactivity was Rutherford's home turf. It had earned him the Nobel Prize in Chemistry in 1908, even though he was not a chemist and even though he regarded chemistry with disdain. (Chapter 5)

Rutherford was not content to pursue scientific knowledge for its own sake; for him, it was a personal race, a competition. His biographer, Arthur Eve, gave an example of this in his book about Rutherford, quoting a letter Rutherford sent to his mother on Jan. 5, 1902:

I am now busy writing up papers for publication and doing fresh work. I have to keep going, as there are always people on my track. I have to publish my present work as rapidly as possible in order to keep in the race. The best sprinters in this road of investigation are Becquerel and the Curies in Paris, who have done a great deal of very important work in the subject of radioactive bodies during the last few years. (Eve, 1939, 80)

In 1922, Rutherford had just completed another major phase of research: follow-up experiments to his 1919 alpha-ray bombardment work. At about the same time that Wendt and Irion announced their results, Rutherford submitted or was about to submit a pair of papers. On May 6 and 13, 1922, *Nature* published the two parts of the Rutherford paper "Artificial Disintegration of the Elements." Wendt also published a two-page article for the general public in *The Nation*. Here's a timeline that summarizes the events:

Timeline of Rutherford-Wendt Publications

March 11, 1922
    Wendt and Irion report results at the American Chemical Society meeting in Evanston, Illinois.
April 1, 1922
    Nature publishes letter to the editor from Rutherford in advance of Wendt and Irion's published paper. (*Science* publishes same letter on April 21)
May 6, 1922
    Nature publishes Part 1 of Rutherford's disintegration paper.
May 8, 1922
    Wendt and Irion submit manuscript for publication to the *Journal of the American Chemical Society*.
May 10, 1922
    The Nation publishes Wendt's two-page article for the lay public.
May 13, 1922
    Nature publishes Part 2 of Rutherford's disintegration paper.
May 25, 1922
    Le Matin publishes article by Charles Nordmann about Rutherford's alleged great alchemical transmutation.
May 26, 1922
    Science publishes Wendt's response to Rutherford's letter.
Sept. 1922
    Journal of the American Chemical Society publishes Wendt and Irion's paper.

In his May 26 letter to the editor, Wendt took Rutherford to task for his unprofessional attempt to make an "intelligent comment" in the absence of relevant facts. What else does this timeline reveal? With one exception, I do not know the submission dates of the publications. Without these facts, I am limited in my ability to make interpretations of the fascinating timing of events. I can, however, offer a few observations.

By March 11, the date of the American Chemical Society meeting, Rutherford either had submitted his papers to *Nature* or, at a minimum, knew that he was soon to do so. In his May 10 article in *The Nation*, Wendt was cordial and respectful to Rutherford. Wendt gave Rutherford credit where it was due, although Wendt did write that he had demonstrated a much larger effect than Rutherford had. In his article in *The Nation*, Wendt made no mention of Rutherford's April 21 letter. Either Wendt's article in *The Nation* went to press before Rutherford's letter, or Wendt ignored Rutherford's hostile comment in *Science*.

Headline from May 25, 1922, news article in Le Matin:
"Professor Rutherford: Modern Alchemist,
HIS DISCOVERIES HAVE ACCOMPLISHED THE
TRANSMUTATION OF THE ELEMENTS."

The 1922 *Le Matin* newspaper article was written by Charles Nordmann, a French astronomer who, in a 1919 guest column in *Le Matin*, had written an exaggerated article about Rutherford's "immense discovery," citing it as the first achievement of the transmutation of elements. Of course, Rutherford had not transmuted any elements in 1919 or even 1922; nor did he even claim he did. (See Chapter 15).

In the shadow of Rutherford's preemptive strike against Wendt and Irion, and a year after Wendt and Irion's paper published, three groups of scientists reported their failures to replicate the exploding-wire experiment, as we shall see next.

CHAPTER 19

Flawless Yet Discredited Experiment

Other Scientists Suggest That Their Own Failures Disprove Wendt and Irion's Successes (1922-1924)

In 1922, Gerald Wendt and Clarence Irion at the University of Chicago reported the successful synthesis of helium from their exploding-wire experiment. In the following years, only a handful of scientists published replication attempts. They all failed to confirm the original claim. On the other hand, none of these groups identified any errors of protocol or analysis in the Wendt and Irion experiment or found that they overlooked a key assumption. Yet the experiment has been recorded by historians as a failure.

In this chapter, I analyze each of these reported failures, I discuss how historians have interpreted these failures, and I summarize Wendt's situation at the close of this series of events.

On March 11, 1922, Wendt and Irion announced their discovery. By May 1922, Wendt and Irion's peers at the University of Chicago had started working on a replication attempt. A news story by Mortimer A. Egan in *Illustrated World* provided the potential significance of the research:

> The custom prevailing in scientific research ... provides that, in the case of all important discoveries, the experiments should be repeated by other workers as a check before being

accepted as conclusive. In this case, a further check on this discovery will be provided by other workers at the University of Chicago who have been making independent investigations. The final verdict will be drawn from all the data thus obtained. But, if, as everything indicates, the results achieved by Wendt and Irion are substantiated, we are facing what is undoubtedly the most important scientific achievement of recent years — an achievement to rank with the discovery of radium and ether waves. (Egan, 1922, 352)

There were three well-known published replication attempts:

Samuel King Allison and William D. Harkins at the Kent Chemical Laboratory, University of Chicago, U.S.
Sinclair Smith, at the Mount Wilson Observatory, under the auspices of the Carnegie Institution of Washington, U.S.
Henry Briscoe, Percy Lucock Robinson and George Edward Stephenson, University of Durham, U.K.

Here is the timeline for the replication attempts, including the original Wendt-Irion paper:

Timeline of Wendt-Irion Paper and Replication Attempts

Sept. 1922
  Journal of the American Chemical Society publishes Wendt and Irion's paper.
Aug. 11, 1923
  Allison et al. submit manuscript to the Journal of American Chemical Society.
Nov. 15, 1923
  Smith submits manuscript to Proceedings of the National Academy of Science.
Jan. 1, 1924
  Proceedings of the National Academy of Science publishes Smith's paper.
April 1924
  Journal of American Chemical Society publishes Allison's paper.
Nov. 24, 1924
  Briscoe submits manuscript to *Journal of the Chemical Society Transactions*.
1925
  Journal of the Chemical Society Transactions publishes Briscoe's paper.

In addition, the Chace-Watson bibliography of the exploding-wire phenomenon suggests that Hantaro Nagaoka and Tetsugoro Futagami, at the Tokyo Institute of Physical and Chemical Research, may also have attempted a replication.

Smith's Replication Failure

Sinclair Smith (1899-1938), a physicist working at the Mount Wilson Observatory, was the first scientist to publish a replication attempt. Smith was a colleague of John August Anderson, who was also affiliated with the Mount Wilson Observatory. It was Anderson's method that Wendt and Irion had used, as they thought, to decompose tungsten into helium. Smith was 24 when he performed the replication attempt and was a Ph.D. candidate in physics at the California Institute of Technology. After his paper published in 1924, Smith went to work at Rutherford's Cavendish Laboratory for a year. (Smith, 1924)

Smith attempted the replication at the request of Anderson. Smith began his paper by reminding readers that, in 1914, physicist George Winchester "had shown that the helium found in [Winchester's experiments was] merely occluded, and not a product of electrode disintegration."

This is not true and reveals Smith's (and presumably Anderson's) bias. As shown in Chapter 13, Winchester wrote that one of his experiments "suddenly, after running for fifteen days, gave out an enormous amount of helium for a few days." He repeated the experiment and again observed new helium and neon. He didn't perform tests to see whether helium was occluded; he just assumed it.

In Smith's paper, just under two pages in length, he explained that he had some difficulties establishing the proper experimental conditions. As far as results, he failed. He found no helium, not even "occluded" helium. His paper lacks detail and depth and does not provide convincing evidence that he made a serious replication attempt. Smith's paper does not attempt to address any possible error by Wendt and Irion. His paper, therefore, has little relevance, if any, to Wendt and Irion's report.

Sinclair Smith

Allison and Harkin's Replication Failure

Samuel King Allison (1900-1965), at the time a 23-year-old chemistry Ph.D. candidate, and William Draper Harkins (1873-1951), at the time a 50-year-old professor of chemistry at the Kent Chemical Laboratory, University of Chicago, performed the next replication attempt. Allison later ventured into the field of physics and briefly, like Smith, studied at Rutherford's Cavendish laboratory. During World War II, Allison worked on the Manhattan project. His co-author, Harkins, later performed research in nuclear chemistry, studying the structure of atoms and helping to develop the hydrogen bomb. (Allison and Harkins 1924)

CHAPTER 19 • 203

*Samuel King Allison (left) and William Draper Harkins (right)
(Photo Credit: National Academies Press)*

Allison and Harkins began their paper discussing atomic theory, focusing on concepts relating to Einstein's theory of special relativity and a concept called "packing effect" that was related to Rutherford's (incorrect) theory of the structure of the atomic nucleus. In their introduction, the authors mentioned nothing about their experimental findings, devoting the entire section to theory. They followed the same unscientific approach — placing theory over experimental fact — used by Rutherford in his letter to *Science*. Allison and Harkins discussed the theoretical effect of transmuting hydrogen to helium and the associated mass deficit that would occur with such a reaction, thereby giving rise to net energy release.

They also wrote about the improbability of helium disintegrating into hydrogen. After outlining the current theoretical constraints, they wrote in their introduction that "there seem to be no good theoretical grounds for supposing that temperature alone will produce the disintegration of atomic nuclei." Wendt and Irion had hypothesized that extremely high temperature, as attained in their experiment, could produce the disintegration of atomic nuclei. Allison and Harkins tried to use theory to cast doubt about Wendt and Irion's experimental results.

Despite their preconceived objections on grounds of theory, Allison and Harkins attempted to repeat the Wendt and Irion experiment. They performed more than 100 discharges, using a variety of conditions and configurations, feeding a 40,000-volt arc through fine metallic wires and a gap in a high vacuum. Electrodes were completely volatized, they wrote, "so that any gasses formed could not remain occluded in them." In 20 of the discharges, they ran experiments that "correspond[ed] to the conditions in the experiments of Wendt and Irion, but no helium was detected."

They wrote a fascinating comment about the lack of detected helium. "In view of the fact that many experimenters have reported that helium is occluded in appreciable quantities in metal electrodes," Allison and Harkins wrote, "it may seem remarkable that no helium at all was discovered in this work. This may be due to the very small quantities of tungsten and platinum volatilized."

The lack of helium is indeed remarkable. One interpretation is that Allison and Harkins effectively provided a second disproof of the occlusion hypothesis. Another possibility is that they did not use enough tungsten or platinum in their experiments to create helium. Allison and Harkins also reported other experiments in their paper: discharges through mercury vapor, and discharges through hydrogen gas. They apparently put great effort into their attempts; however, they detected no helium.

Allison and Harkins, in their conclusion, offered a theoretical explanation for their failure to detect helium. "The theory of atomic disintegration is discussed," the authors wrote, "and it is pointed out that the failure of the electrical discharge methods to disintegrate the atom is due to the fact that energy is not transmitted to the nucleus in a sufficiently high concentration. Electrons with velocities of several million volts may prove effective, though thus far only high-speed alpha particles have induced disintegration, as in the experiments of Rutherford."

In their assessment of possible causes of their failure to replicate Wendt and Irion, they did not, however, say that they may have failed to properly replicate the required conditions. Their paper also did not try to address any possible error or assumption by Wendt and Irion. It was simply a failure to replicate and has little, if any bearing, on the original work.

Mat Nieuwenhoven, an editor of this book, analyzed Allison and Harkins' paper. Nieuwenhoven was most interested in the choice of wire(s) used by the two groups and the switching mechanism. Nieuwenhoven learned that Wendt and Irion used a single uninterrupted wire for their explosion target and an external spark gap to close the circuit. Allison and Harkins used two wires separated by a small gap and a mechanical switch to close the circuit.

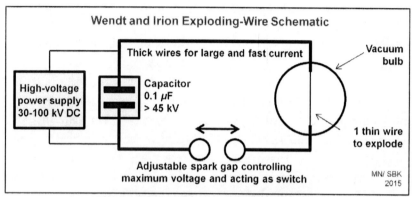

Wendt and Irion Exploding-Wire Schematic

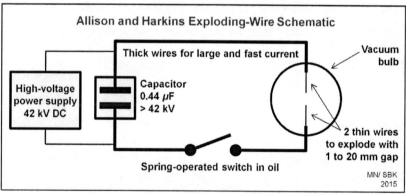

Allison and Harkins Exploding-Wire Schematic

These are crucial differences. The resistance in the air gap stays extremely high until the moment the gap is sparked over. A strong current through air (for example, lightning) ionizes the air, and the resistance drops greatly. It's like closing a switch, in this case extremely rapidly. In the Wendt and Irion design, the current from the capacitor can then go rapidly to and through the single wire to be exploded. This is such a radical change to the circuit that it cannot be called a valid replication attempt.

Briscoe's Replication Failure

The next replication attempt was performed by Henry Vincent Aird Briscoe (1888-1961), Percy Lucock Robinson (1891-1973), and George Edward Stephenson (1892-1965), at Armstrong College, University of Durham, United Kingdom. (Briscoe, Robinson, and Stephenson, 1925)

Briscoe, a professor of inorganic chemistry, was 36 at the time. He earned his Ph.D. in 1909 at Imperial College, London at 21, then took up a teaching post at Armstrong College, University of Durham. In 1932, he went back to Imperial College for 22 years, first as the chair of the Organic Chemistry Department and, in 1949, as the head of the Chemistry Department. Briscoe was also an adviser to the British Security Service, MI5, during and between the world wars.

Robinson had studied under Briscoe and earned his master's of science degree in chemistry in 1918. He collaborated with Briscoe from 1921 until 1932 and was 33 when they tried to replicate Wendt and Irion's work in 1924. After 1932, Robinson went to Durham University, Newcastle, to teach chemistry. I have been unable to locate any biographical information about Stephenson.

The Briscoe group's replication attempt was funded primarily by a grant from the British Chemical Society, and the organization published the group's paper, as well. The society's dual roles of fiscal sponsor and publisher diminishes the credibility of the publication. Based on the explanations provided in the paper, the Briscoe group had great difficulty replicating the correct conditions.

"All attempts to seal heavy tungsten wires into Pyrex glass," the authors wrote, "as described by Wendt and Irion, failed completely, but, fortunately, it was found that molybdenum sealed fairly well into Durosil glass and could, unlike tungsten, be obtained as drawn wire of the heavy section required to carry the discharge without material loss."

They also obtained samples of fine tungsten wire but had difficulty with these, too. "It was quite impossible to spring these wires unsupported between the electrodes, as Wendt and Irion stated they had done," the authors wrote. "With much patience and luck, crystalline filaments from an old drawn-wire lamp could be fitted in this way, but they were so fragile as to be useless in practice."

Henry Vincent Aird Briscoe

The most striking concern about the Briscoe group's paper is that the researchers did not obtain the well-known optical and acoustic signatures of detonation. "The air-gap was screened from sight," the authors wrote, "and the explosion was well seen by four observers: It was attended by a dull thud in the bulb, and by a bright flash in which the whole of the filament was seen to be involved. Both the noise and the brightness of the flash were much less than we had anticipated."

The lack of these fundamental characteristics, with little uncertainty, indicates that the authors did not achieve the required conditions. Despite these inadequacies, the authors wrote, "It will be remarked that these effects differ substantially from those described by Wendt and Irion: We nevertheless regarded this explosion as complete and satisfactory. ... We conclude, therefore, that our experiments afford no evidence that tungsten can be made to yield helium when exploded by an electrical discharge."

The Briscoe researchers asserted that the failures of Smith and, separately, of Allison and Harkins lent support to their failure. After devoting half a page to summarizing those two papers, the Briscoe group left readers with this conclusion: "Hence, it seems clear that the statements made by Wendt and Irion must be attributed to erroneous observations and have no foundation in fact."

Ironically, the Briscoe group's comment, despite the researchers' self-confidence, had no foundation in fact or science. Not one of these groups, including Rutherford, identified any error of experimental protocol, analysis, or assumption by Wendt and Irion. Three failed replications (or even three dozen) do not mean that an original set of experiments is wrong.

Nuclear effects in exploding wires were not independently confirmed until 2002 at the Kurchatov Institute in Russia. (Urutskoev, 2002)

The Failure of History

The Briscoe group's conclusion, despite its lack of foundation, was recorded as the final word on the matter by the science media. *Science Abstracts: Physics,* a service that summarized and broadly disseminated key research, pulled a few sentences out of the Briscoe paper, slightly edited them, and left the reputation of Wendt and Irion's discovery in the hands of Briscoe:

> These phenomena are attributable to the momentary attainment of an exceptionally high temperature in the substance of the wire. In 1922, Wendt and Irion reported the results of experiments on the explosion of fine tungsten wires by Anderson's method. When the explosions were carried out in a vacuum, some gas was formed, and this, subjected to an ordinary high-tension discharge, gave a spectrum in which the yellow helium line, D_3, was

consistently observed. ...

It is concluded that these experiments afford no evidence that tungsten can be made to yield helium when exploded by an electrical discharge. ... The conclusion is, however, supported by experiments of S. Smith and those of Harkins and Allison, and so it would appear that the statement of Wendt and Irion as to the formation of helium has no foundation in fact. (*Science Abstracts: Physics*, 1925, 392)

The abstracts were published, compiled, and edited by the British Institution of Electrical Engineers in association with the British Physical Society of London. The following year, in 1926, reviewers for the U.S. National Research Council, under the auspices of the National Academy of Sciences, published their conclusion. The reviewers briefly described the Wendt-Irion claim and followed it with Rutherford's assertion that the Wendt-Irion claim was wrong because it contradicted Rutherford's understanding of nuclear theory.

The reviewers wrote that "Smith obtained negative results in an attempt to check their findings." This is wrong. Smith made no attempt to check Wendt and Irion's findings. He made a half-hearted attempt to replicate their findings and failed. The National Research Council reviewers concluded that "the disintegration of tungsten vapor by the electric discharge must be regarded as not yet proved." They were correct about that. (Kovarik and McKeehan, 1926, 133)

In a 1930 book, Rutherford and two of his colleagues took the opportunity to put the final nail in the coffin. "The experiments of Wendt and Irion," the authors wrote, "were repeated very carefully by Allison and Harkins, but no evidence of the production of helium was found." (Rutherford, Chadwick, Ellis, 1930, 315)

Rutherford was talking, to use his own famous word about the prospects of atomic energy, "moonshine." Failure to replicate means failure to replicate, nothing more. Additionally, Rutherford's reference to Allison and Harkins was self-serving, because their paper cited Rutherford as finding the only proven method so far for man-made disintegration of atoms.

An Unwinnable Battle

With such obvious deficiencies in the aforementioned replication attempts and in Rutherford's unscientific critique, it is reasonable to ask why Wendt did not fight back. I've not been able to find any direct answer to this, but I can offer the following information. Immediately following the completion of their work, Wendt came down with an illness and was required to rest for a year. Irion left the University of Chicago to continue his studies elsewhere.

I suspect, however, there were three more factors, far more significant than these logistical challenges. First, Wendt knew what he and Irion had done, and he knew what they had accomplished. He knew that none of the critics had identified any errors of protocol or analysis or found that they overlooked a key assumption. Nor had Rutherford, Smith, the Allison group, or the Briscoe group revealed any potential flaw in or concern with his and Irion's work.

Wendt knew there was nothing left for him and Irion to defend. Wendt, perhaps more than anyone, as a prolific scientific communicator and as a teacher of science teachers, was intimately familiar with the fundamental concepts of scientific research and the scientific method. He knew that he and Irion had done their best and the rest was now up to others in the scientific community.

Second, he knew that Rutherford and the failed replicators had gone beyond the scientific method in their critique of his and Irion's work. Wendt knew that he had called Rutherford out on his poor behavior. I suspect that Wendt realized the practical limitations that he faced, and he accepted that he could not win this battle, which was less scientific than political.

Third, Wendt's greatest passion and interest were not discovery, fame, or Nobel prizes. His joy was to share science's wonders with the world and to see science used for the benefit of mankind. Here are a few words from the dust cover of Wendt's book *Science for the World of Tomorrow:*

As humanity moves ever forward from the cave to the house, from clan to nation, from slavery to leisure, the centuries show that, in one respect, progress is clear. Man's outstanding achievement has been in the use of his intelligence to improve the environment of human life. This is science.

[I do] not consider science merely as a great body of knowledge, a vast encyclopedia of facts and principles. Neither [do I] regard it as a great collection of useful instruments, of gadgets and machines, as engineers and businessmen often think of it. Both of these are only parts of the story, for essentially science is a method of solving problems, and the problems are essentially those of human beings. (Wendt, 1939)

CHAPTER 20

Accidental Alchemist

Miethe Finds Trace Amounts of Gold in the Residue of Mercury Vapor Photography Lamps (1924-1926)

At least as early as January 1922, rumors of transmutations of inexpensive materials to gold were appearing in newspapers around the world. The idea of transmuting lead to gold wasn't entirely illogical at the time. After all, if Rutherford was correct, as people assumed, that he knocked a proton out of a nitrogen nucleus, couldn't someone knock three protons out of lead and end up with gold?

Rutherford proved that an atom could be shattered and that hydrogen particles were emitted. He did not, however, observe and report a transmutation, so the holy grail of transmutation was still up for grabs.

For men and women of science, the idea of transmuting lead to gold was more of a philosophical quest than a practical objective. They also realized that, according to Einstein's theory of special relativity, a conversion of elements resulting in a mass deficit could release an extraordinary amount of energy, and this quest for energy drove the practical inquiry for atomic energy. Finally, in 1924, two scientists reported the transmutation of a base element into gold.

Utopia Through Transmutations

The series of events that I'm about to report was preceded by a forward-looking July 31, 1923, article in the *New York Times* which quoted Paul Darwin Foote (1888-1971), a scientist with the National

Bureau of Standards. Foote gave a lecture at Columbia University's Department of Physics the day before and explained how the energy released by the conversion of elements might someday serve the energy needs of the entire world.

"The older alchemists desired to create gold," Foote said. "The modern alchemists would [attempt to] destroy it, for the energy from the destruction of gold is immensely valuable. One gram of gold destroyed by transformation into electric energy would be worth $2,600,000 at the present price of electricity."

Foote also said that the governments of the world, with economies based on gold, would face a financial revolution and need to demonetize the metal and make a new standard. Foote was optimistic that a goods- or market-based monetary system would lead to utopia and that "humanity will be emancipated by the scientist."

> **PREDICTS A UTOPIA WHEN ATOMS WORK**
>
> Dr. Foote Says Energy Released in Transmutation May Some Day Serve Whole World.
>
> **EXPLAINS MODERN ALCHEMY**
>
> Declares Changing One Element Into Another Is Today Only a Quantitative Problem.

Headline from July 31, 1923, New York Times *article*

Foote was an authority in science and had an illustrious and varied career. He worked at the National Bureau of Standards from 1911 to 1927. For the next two years, he worked for the Mellon Institute of Industrial Research, and from 1929 to 1953, he worked for the Gulf Oil

Corporation, starting as a researcher and ending as a vice president. In his retirement, he was appointed assistant secretary of defense for research and engineering by President Eisenhower. At the time, he was also a special assistant to the secretary of state for atomic energy affairs, and he later served as a member of the industrial advisory board of the Atomic Energy Commission.

Paul Darwin Foote, Photo Credit: AIP Emilio Segrè Visual Archives, Physics Today *Collection*

A few months later, Foote wrote a fascinating article in the September 1924 issue of *Scientific Monthly*, a publication of the American Association for the Advancement of Science. His article began with a historical perspective on alchemy, then deftly revealed how the modern explorations into atomic science were related to the ancient philosophies of the alchemists. Most striking in his article was his understanding (presumably shared by other scientists of the day) of the mass deficit when transmuting elements upward on the periodic table:

> The helium atom may be synthetically constructed from four hydrogen atoms. ... So we know that the atomic weight of helium is 4.00, while the atomic weight of hydrogen is 1.0077. Hence, four separate atoms of hydrogen weigh

4.031, or 0.031 units more than when they are compressed together to form an atom of helium. Thus, the formation of one atom of helium annihilates 0.031 units of mass and, by the general principle of Einstein, results in the creation of [mass times the speed of light squared] units of energy. ... That is, if the hydrogen in two teaspoons of water be converted into helium, 200,000 kilowatt hours of energy is set free, representing $20,000 worth of electrical current or ten thousand dollars to the teaspoonful. (Foote, 1924, 260)

Foote, like other scientists at the time, assumed that all elements were made of hydrogen particles. In some sense, they were correct; today, we would say that all elements are made of protons plus neutrons and electrons. The neutron, as well as the nuclear processes of fission, fusion and neutron capture, was yet to be discovered. Even with the limited knowledge then, Foote and other scientists clearly and correctly understood the energy potential of transmutation. Despite the Utopian angle taken by the *New York Times*, Foote's article in *Scientific Monthly* included his belief that, even if atomic energy obliterated poverty and suffering, new problems likely would appear.

Very Expensive Gold

Scientists were aware at the time that there was no cost-effective method of making transmutations. A brief news blurb circulating the United States from one of the wire services reflected this view. The *Wichita Daily Times* reported it on Jan. 27, 1922: "Transmutation of metals was universally accepted as impossible a few years ago. Now, some scientists say they'll soon be turning lead into gold, though the cost may be prohibitive."

Between 1924 and 1926, two scientists, one in Germany and another in Japan, were independently performing experiments unrelated to transmutation. To their great surprise, they detected traces of gold in their materials where none had existed before the experiments. I do not have sufficient access to and understanding of these German and

Japanese papers, nor access to any additional critical papers in those languages that would enable me to perform in-depth analyses of these claims. Therefore, I am not in a position to offer my own opinion on the credibility of these claims.

However, the German and Japanese claims are interesting for the stories alone. The news accounts, particularly those in the journal *Nature*, provide useful perspective. This and the next chapter provide a trail of breadcrumbs for future historians, particularly those with skills in German and Japanese, to unearth the complete story.

I will, for the most part, present the stories of the German and the Japanese scientists chronologically. However, there's a slight overlap between the two narratives, so from time to time, their stories will be intertwined. Both the German and the Japanese scientists claimed that they had transmuted mercury to gold. This was just a single step down the periodic table, rather than transmuting lead to gold, which was three steps.

The logic was not preposterous at the time. If Rutherford could knock off a single proton from nitrogen, why couldn't the same be done with mercury? Rutherford's hypothesis about knocking a proton off nitrogen was wrong; however, that was not revealed until 1925, and even then, physicists were slow to recognize that Rutherford's original hypothesis was wrong. But in 1924, unintentionally thanks to Rutherford, the idea that mercury could lose a proton and become gold was thought to be, at least partially, theoretically feasible. Scientists still had concerns that transmutation of any significant mass would be very difficult, considering the high energy required to break apart atoms.

Gold-Laced Sooty Deposit

The German transmutation claim came from the Berlin laboratory of chemist Adolf Miethe (1862-1927), 62 at the time, a pioneer best known for his work in photography. Miethe was well-known for inventing a three-color photographic process and, along with Johannes Gaedicke, inventing the magnesium photo flash. Miethe was famous in the moving-picture industry for his improvements on camera lenses and his three-color process. Among other professional activities, he was a professor of photography, photochemistry and spectral analysis at the Technische Hochschule (an engineering college, not a high school) in Charlottenburg, Berlin.

Miethe was assisted by chemist Hans Stammreich (1902-1969), 22 at the time, who studied under his direction. Stammreich was born in Brazil and was a pioneer of Raman spectroscopy.

Miethe published his first paper on his quicksilver (an older name for mercury) -to-gold transmutation in July 1924 in *Naturwissenschaften*. He was doing experiments with mercury vapor lamps for use as a source of ultra-violet rays. After a time, the lamps stopped working. On inspection, he noticed that a sooty deposit formed in the quartz tubes. Miethe tested these deposits and, to his great surprise, detected gold. (Miethe, 1924)

Adolf Miethe

Hans Stammreich

One of the wire services picked up the news of the *Naturwissenschaften* paper and ran a short story on it a few days later. Here is the *Iowa City Press-Citizen* publication of the wire story, published on July 30, 1924. The public quickly saw the immediate potential economic impact of Miethe's discovery:

> MODERN ALCHEMY — Gold has been pouring into our country from abroad at a rate of $40 million a month, so far this year. We now have 4, 500 million worth, or more than half of the world's supply. It helps keep prices high. Our predicament will be serious if other leading nations, drained of their gold, quit using the yellow metal as the basis of their money systems and instead use a more scientific backing

such as units of human energy. Dr. Adolf Miethe, German, claims he has discovered how to turn mercury into gold. Cost is above value, so far. What if a cheap process is found? Turning lead into gold is not impossible. What a ghastly joke if, having cornered most of the world's gold, it suddenly were made worthless by chemical discovery.

Alchemy historian Robert Nelson wrote a concise explanation of Miethe's work in Nelson's book *Adept Alchemy*:

> In July of 1924, Drs. Miethe and Stammreich announced that they had changed mercury into gold in a high-tension mercury vapor lamp. The experiment produced $1 worth of gold at a cost of $60,000. [It ran] for 20-200 hours. The lamp consumed 400-2,000 watts. ... The yield of gold was minute: 0.1-0.01 mg. The mercury and the electrodes were analyzed and determined to be free of gold before the experiments.

On Aug. 9, 1924, *Nature* published a review of the July 18, 1924, *Naturwissenschaften* paper that provides the technical details:

> Some thirteen months ago, professor Miethe adopted an improved type of mercury-vapour lamp, made by A. Jaenicke, for his experiments on the coloration of transparent minerals and glasses induced by ultraviolet rays.
>
> Early this year, [Miethe] and his assistant, H. Stammreich, observed that when the current employed was too strong, the character of the emitted rays soon changed, and that a black deposit formed inside the lamp. (It is not stated if the lamp was of quartz or of glass.) This observation was confirmed by Jaenicke, who stated that, on distilling the mercury recovered from old lamps of the improved type, he had found residues which he could not identify. He supplied Miethe with about 0.5 gram of the residues (which had been obtained from 5 kg of mercury), and the latter, after careful investigation, discovered

that they contained gold, in addition to other substances which were undoubtedly present as impurities in the original mercury.

According to Jaenicke, all the mercury used in the lamps had been twice distilled before use. In his successful experiments, Miethe always used a potential difference of 170 volts between the electrodes, which were in direct communication with the air outside [the cell]; the current was passed for 20-200 hours; and the lamp consumed 400-2,000 watts. It appeared probable that a minimum potential difference is essential, for no trace of gold was found in mercury lamps of the old type that had been long in use, and negative results were also obtained with lamps of the improved type when the potential difference was below a certain figure.

Owing to the minute quantity of gold obtained, namely, 0.1-0.01 mg, special precautions had to be taken in identifying it. In every test, an amount of the original mercury equal to that which was removed from the lamp was analyzed and found to be free from gold both by Miethe and independently by K. A. Hofmann; and it was proved that no gold was present in the electrical connections. Further, the very delicate analytical methods elaborated by Haber failed to show with certainty the pre-existence of gold. When the residue left, after distilling off *in vacuo* the mercury from the black deposit, was treated with nitric acid, there remained yellow well-formed crystals, cubical and octahedral in form, and with a highly reflecting surface; and when the mercury was removed from the deposit by volatilizing it at a red heat, the resulting substance was bright gold in color and of reniform or botryoidal shape.

The substance was found to be malleable, it gave the streak of fine gold and the characteristic color when the polished film was observed by doubly reflected light. It was easily soluble in aqua regia, and on evaporation the solution gave crystals precisely similar to those obtained when ordinary gold is so treated. Identical results were also obtained by the purple-of-

Cassius test. It was not possible to make an atomic-weight determination or to attempt to prove the production of helium, hydrogen, alpha or beta rays. (*Nature*, 1924, 197-98)

Miethe and his colleagues seemed to have unambiguously identified gold. The article also stated that two people independently verified that Miethe's gold did not exist in the mercury before the experiments. The article stated that one of the analysts checked for pre-existing gold in the electrical connections, but it did not state whether the electrodes were checked for pre-existing gold.

The next news article that got my attention was published on Oct. 19, 1924. It gave an outline and fair analysis of the Miethe experiment. The author was confident of the direction in which the science was moving and the implications for world finance:

Financiers say that [artificial gold] would wreck the financial structure of the world. If gold became suddenly worthless, the collapse of world credits would be 1 million times greater, they fear, then all the disastrous effects of the great war. Very well. Let the financiers get ready. For artificial gold is one of the things that science is going to produce. Perhaps not this month, or this year, or even in 10 years. But someday it will be done. And probably it will be done cheaply. (*Oakland Tribune*, 1924)

There is no byline on the article, but there are three initials at the end of the article, E.E.F. The author was almost certainly Edward E. Free, the editor of *Scientific American* magazine. A month later, *Scientific American* announced that it was going to sponsor a replication attempt of the Miethe-Stammreich transmutation claim. The magazine issued a press release on Nov. 11, 1924, according to an Associated Press news story printed in the *Bakersfield Morning Echo*. The newspaper's editors chose a dramatic headline — ALCHEMIST OF FABLE MAY BE PUT TO SHAME. The news story gives a general outline of the experiment and the objective of the replication, which was to be performed at New York University.

The scientific replication attempt was depicted as a horse race.

> **ALCHEMIST OF FABLE MAY BE PUT TO SHAME**
>
> **TO TEST PLAN OF EUROPEAN SAVANT**
>
> Experiments Will Be Conducted in New York College
>
> **PROCESS IS COSTLY**
>
> Reputed Discovery of German Declared to Be Accident

Nov. 11, 1924, Associated Press story in the Bakersfield Morning Echo *announcing the forthcoming replication attempt at New York University*

Another Associated Press story from that time mentioned that the New York University physicists were planning to use a different method from the one originally used by Miethe. A news article from Nov. 23, 1924, confirmed this. The newspaper reported, among other things, that the New York University physicists did not intend to follow Miethe's protocol:

> They have simpler equipment than heretofore used. They hope to produce gold at a greatly lessened cost. Miethe used the ultraviolet ray in working upon the quicksilver. The New York scientists will apply a different light for the purpose of cracking the atoms of quicksilver. ... Taking what the German scientist discovered as a base, the Americans have been seeking ways to cut down the cost of operation and believe they are near the goal. (*Charleston Daily*, 1924)

There were two intrinsic problems with the New York University physicists' idea to "improve" the protocol before they had confirmed that they could make it work with the original protocol. First, they were trying to perform an experiment on the leading edge of science.

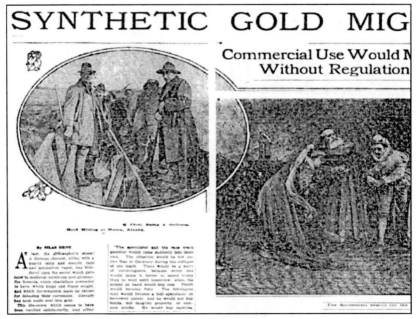

Full-page news headline from Nov. 23, 1924, article in New York Times. *Captions, above, left to right: Gold Mining at Nome, Alaska; The Alchemists Search for the Secret of Making Gold; Moving Gold in New York Streets*

They could not be expected to fully understand all of the specific aspects of a novel experiment, least of all one that they had no part in developing. They could very easily and unknowingly have varied a key parameter or protocol and, assuming good faith, unintentionally have caused the experiment to fail. The second reason is that, by intentionally varying the experiment, they left themselves open for criticism that they did not correctly replicate the conditions of the experiment.

Disrupt the World

The same day, the *New York Times* splashed a dramatic headline across a full page in an article titled "SYNTHETIC GOLD MIGIIT DISRUPT WORLD: Commercial Use Would Mean Chaos and Finance Without Regulation, Economist Says."

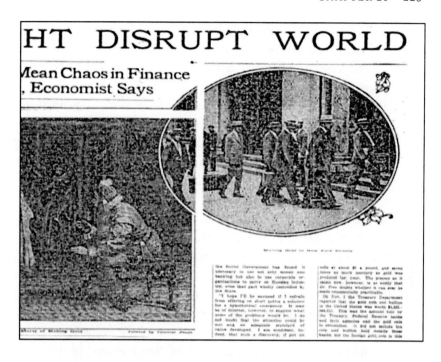

The *Times* relied on the verifications reported by *Nature* that the Miethe claim had been independently checked, though not independently replicated:

> The discovery, which seems to have been verified satisfactorily, may affect every man's life. A battery of [scientists] in New York is making a series of tests to see whether the thing can be commercialized. As the German chemist did it, about $2 million worth of electric current would be required to manufacture about $300 worth of gold. But suppose the metal could be produced at half or one-hundredth of the present cost of mining and refining it. What effect would it have on the financial structure of the world? What would it do to the monetary systems founded on gold? What would happen to contracts, to stocks and bonds and debts? (*New York Times*, 1924)

The *Times* spoke with Benjamin M. Anderson Jr., an economist with Chase National Bank of New York, who insisted that those questions were purely hypothetical and that the possibility of cost-effective artificial gold was too remote.

The *Times* quoted Anderson indirectly. "If such a thing were to happen," he said, "the principal nations of the world might enter into a covenant to prevent the artificial production of gold, at least until they could cast about for a new basis of exchange and readjust their monetary systems to it." The *Times* then quoted him directly:

> Any such revolutionary discovery as this, if commercialized, would turn the whole field of technology upside down. If baser metals could be transmuted into gold, then technologists would learn how to transmute other metals, and probably [learn] how to release power from the atom. Apart from the disturbances to values due to the changed relation of gold to goods, there would be enormous disturbances as between the values of commodities. Some would become cheaper as compared with others. There would be the biggest financial and industrial readjustment the world has ever seen. ... A tremendous drop in the value of gold as compared with goods, and of dollars as compared with goods, would ruin creditors and release millions of debtors. (*New York Times*, 1924)

A Magazine's Public Duty

In December 1924, *Scientific American* formally announced in a press release its intention to sponsor the Miethe replication attempt at New York University. Soon after the press release, the magazine published an editorial titled "Why We Are Trying to Make Gold" that tried to explain its motives.

CHAPTER 20 • 227

> ### What We Hope To Do
>
> Professor Adolf Miethe, of Berlin, reports that he has converted quicksilver into gold. The SCIENTIFIC AMERICAN is now arranging for the repetition of this experiment.
>
> If the transformation of quicksilver into gold is confirmed, we will endeavor to discover exactly what conditions control the conversion.
>
> We will obtain data for an estimate of the cost.
>
> We will make the best possible estimate of the time—if ever—when the world will have to face the threat of cheap gold.
>
> We will communicate the results of this investigation to the public unless, as a matter of public policy, it seems wise to delay their publication.

December 1924, Scientific American's *stated intentions*

The article began by considering how revolutionary — in the face of 300 years of failed alchemy — the Miethe-Stammreich transmutation claim would be, if valid:

> Where the labor and intelligence of so many thousands of alchemists came only to failure, a German college professor seems to have succeeded by absolute luck. His magic wand is not any mystic stone of the philosophers; it is the more modern magic of electricity. This, at least, is what most of the scientists of the world are now inclined to believe as a result of [these] remarkable observations.
>
> Professor Miethe was not searching for the philosopher's stone; far from it. He was engaged in researches on ultra-violet light. To procure this light in sufficient intensity, he employed a quartz lamp containing the vapor of mercury, the liquid metal familiar to everyone under its other name of quicksilver. The vapor of mercury was contained in a bulb made of fused quartz. Through this

vapor, there was passed a powerful current of electricity. After a time, professor Miethe noticed that the inside of the quartz bulb had become coated with a black film. He analyzed this film. It proved to contain gold.

This was astonishing, but it might be accidental. The professor started afresh. He got some new mercury and carefully purified it. He divided it into two portions. One portion he placed in his quartz lamp and then ran the lamp for two hundred hours. On analysis, gold was found in the mercury that had been in the lamp; no gold was found in the other portion of the same mercury, the portion that had not been exposed to the electric current. The lamp itself was then taken apart and analyzed. No gold was found in any part of it. The conclusion was inevitable that a very little gold had been created somehow during the operation of the lamp. Presumably, it came from the conversion of a few of the mercury atoms into atoms of gold.

That is where the matter stands as these paragraphs are written. Admittedly, the experiment needs confirmation. But Professor Miethe is a scientist of distinction. His precautions seem to be adequate. There is not the slightest question of his honesty. On the contrary, there is every reason to believe that he has actually done what he thinks that he has done. It is probable that the secret of the philosopher's stone has been found.

If this is true, it is one of the most important scientific discoveries ever made, and not all of its importance is scientific. Gold is not only beautiful and scarce and valuable; it is also the basis of modern finance. (*Scientific American*, 1924)

Scientific American then gave a brief explanation of the gold standard and explained that the foundation of modern currencies rested precariously on the rare metal. The experiment was going to be watched closely by every banker, investor, and economist in the nation and possibly internationally. The magazine explained the key concern:

> [Miethe's] process is not cheap. It costs him thousands of times as much to make his gold from quicksilver as it would to go and buy some gold in the nearest jewelry store.
>
> But this expensiveness of the method may not be inevitable. Most things cost a great deal when they are first discovered. The original electric lamp cost, it is reported, more than a thousand dollars. Now you can buy a better one for thirty cents. (*Scientific American*, 1924)

Scientific American emphasized that, if financial planners could have advance notice about the potential loss of gold's value, the magazine wrote, they would have time to shift the basis of world credits to some other measure:

> The great need of the present situation, therefore, is more facts. Does the Miethe process really make gold? Exactly what does it cost? What are the prospects of cheapening it? Can we expect ever to make gold more cheaply than we can mine the natural gold from the earth? How soon, if ever, can we expect such cheap gold to be available?
>
> Given these facts, the governments and financiers of the world can prepare for whatever is probable. They can take steps to control the gold-making process, if that is possible. Or they can prepare the structure of finance for an easy transition to whatever other standard of values seems necessary or most desirable. (*Scientific American*, 1924)

On the Shoulders of a Lad

For these reasons, *Scientific American* explained, it had funded a replication attempt at an American laboratory. The magazine made arrangements with physicist Harold Horton Sheldon (1893-1964) of Washington Square College of New York University to use its facilities. Sheldon was born in Brockville, Ontario, Canada, and earned his Ph.D.

at the University of Chicago in 1920. Sheldon, 31 at the time, was an associate professor at New York University in 1924 and became a full professor in 1927.

The experiments were performed by Roger Shepard Estey (1903-1987), who, according to *Scientific American,* was "an experienced physicist and research investigator" and was hired to conduct the work under the direction of Sheldon. *Scientific American* did not mention that Estey was 21 and a master's degree candidate at the time he performed the experiment.

Harold Horton Sheldon, on left; Roger Estey, on right. Edward E. Free, editor of Scientific American, *in the center. (Photo: Scientific American)*

The magazine explained its motives further:

> Mere artificial gold would have tremendous scientific interest; it might be useful in many ways in scientific experimentation; unquestionably the process would give us deeper insight into the many unsolved problems of atomic structure. But these things, interesting as they are, are not the impelling motive of the *Scientific American.* What does impel us is the duty of foreseeing, so far as may be possible, what is likely to happen to gold as a medium of exchange and a standard of financial values. That is what we hope to discover, what we hope to be able to make public in future

issues or to communicate — if it seems that secrecy is necessary — to the leaders and directors of the world's finance.

A sudden debacle of values based on gold would be a world misfortune beside which wars and earthquakes and the Black Death are mere mild annoyances. The one thing that will surely prevent such a debacle is advance knowledge of its imminence. That advance knowledge, the *Scientific American* believes, is its duty to provide, if it be humanly possible to do so. (*Scientific American*, 1924)

The replication attempt began, and, presumably, the future stability of the developed world rested on the shoulders of one 21-year-old college student. Lots happened before the *Scientific American*-sponsored experiment ended and was reported as a failure.

Possible Stammreich Replication

From what I can tell of a Dec. 19, 1924, news story written by Edwin E. Slosson, for *Science Service*, a Washington, D.C.-based news service, Miethe's assistant, Stammreich, performed a partially independent and successful replication of the experiment. I do not, however, have any information about a published journal article, so I can't place confidence in the scientific validity of this news. Here's what Slosson wrote:

> The transmutation of mercury into gold, which was recently reported to have been accomplished by professor A. Miethe of Berlin, and generally questioned by other chemists, is now announced to have been confirmed by his collaborator, Dr. H. Stammreich. He claims to have repeated Miethe's experiment, using mercury, which careful analysis had shown to be "absolutely free from any trace of gold," but which, at the end of the process, was found to contain appreciable amounts of the precious metal. (*Science Service*, 1924)

Slosson quoted William Draper Harkins, a chemist at Kent Laboratory of the University of Chicago, who was reluctant to accept the claim because of his theoretical preconceptions:

> [Harkins] declares that, in the mercury vapor lamp, with which professor Miethe claims to have produced gold, the energy brought to bear upon the atoms of mercury is exceedingly small in comparison with the amounts of energy in all actual artificial disintegrations thus far accomplished.
>
> "According to accepted theories, it also seems probable," Harkins says, "that such small amounts of energy would not be able to penetrate the outside of the atom and get at the nucleus at all. Professor Haber found gold in Miethe's mercury, and this is undoubtedly accurate, but Haber disclaims all knowledge of how the [gold] got in. Experts in this field will not trust any reports of atomic disintegration by large or small currents, unless voltages of millions of volts have been used, until they are supported by experiments of work carried out with the most extreme precautions in such a way as to give definite evidence that the results claimed had been obtained. It is possible that Miethe has such evidence. I have not repeated his experiments. Mercury would be converted into gold if a hydrogen nucleus were lost from, or an electron added to, mercury's nucleus. I have bombarded argon nuclei by helium nuclei within energy corresponding to 5,000,000 volts without their disintegration. The voltage used in a mercury vapor lamp is [too] small." (*Science Service*, 1924)

Two months later, on Feb. 16, 1925, the *Berkeley Daily Gazette* published another article by *Science Service*. Here is an excerpt:

> The quartz [glass], the iron and the carbons of the lamp were also analyzed and pronounced gold-free. Miethe sent samples of these and of the mercury, before and after using in the lamp, to professor Haber, the inventor of the Haber

process for fixing nitrogen, who has been interested in the extraction of gold from seawater and had developed a very delicate method of estimating gold in a minute amount. He reported finding gold, in some cases silver, in the samples that came from the lamps. The amount varied from 1 to 52 parts in 1 billion parts of mercury. From these experiments, ... he concludes that some of the [mercury atoms] have been crumbled away by the electric current passing through the vapor, leaving gold as a residue. (*Science Service*, 1925, Chats)

Estey Begins His Work

In March 1925, *Scientific American* published a one-page article, containing mostly photos, and reported that the experiments were under way at New York University.

Benchtop apparatus of Sheldon and Estey's mercury-to-gold replication attempt (Photo: Scientific American*)*

Possible Replication in Chicago

The *Scientific American*-sponsored replication attempt finished by November 1925. But I found other strange news reports from March and April that year. Some, possibly all, of them were distributed by the wire services. Unfortunately, I have no other information about the following events except what I found in these three brief news reports.

> Chicago, March 26 (AP) — Two machines pronounced among the most efficient thus far, built for the separation of an element into different substances, have been completed in the Kent chemical laboratory of the University of Chicago. One is used for chlorine separation. From the other, mercury is caused to run out a different substance from that which was put in, the lighter substance in the mercury being separated from the heavier by causing the atoms of mercury to jump about in the machine at seven times the speed of the fastest express train. The mercury separation, chemists say, will be of value in connection with the experiments being undertaken on the transformation of mercury into gold. (*Emporia Daily Gazette*, 1925)
>
> Chicago, March 30 — All chemists in several countries are close to their age-old quest — the transmutation of the baser metals into gold — according to an announcement made today by scientists at the University of Chicago. Mercury is the metal used in the experiments. Workers at the Kent chemical laboratory at the university are experimenting on a new machine, one just invented, which is said to be extremely efficient. They already have been able, by its use, to change the weight of the mercury put into the machine, but so far are still baffled in their efforts to change its appearance. The mercury now has the weight of gold, though it does not yet look like it. (*Adrian Daily Telegram*, 1925)

Chicago, April 13 — The statement issued at Kent laboratory, after referring to "recent reports in the press" which indicate that Miethe, in Germany, and Nagaoka, in Japan, "believe they have converted mercury into gold by the use of large currents in a mercury arc lamp," reads as follows:

"Work has been begun in this laboratory on the method by means of which electrons with thousands of times higher velocities are shot into mercury in order to attach themselves to the mercury nucleus and thus produce gold. It is the opinion of those who have begun this work that even the greatest concentrations of energy will be insufficient, and that still more powerful and expensive sources of energy may be needed to be applied." (*Brandon Daily Sun*, 1925)

As a point of reference, gold has an atomic mass of 196.96. Platinum has four dominant naturally occurring isotopes; one of them is platinum-198 with a weight of approximately 197.96. Platinum also has a rare isotope with the weight of about 196.96. That isotope is unstable, and it has a short half-life of nearly 20 hours. Visually, platinum could easily be mistaken for silver. If a scientific paper exists for this experiment, it could have great significance. I contacted an archivist at the University of Chicago, and unfortunately the archivist was unable to find any further details, papers or announcements. It is a tantalizing piece of news considering the fact that Miethe claimed that he had observed silver in addition to gold.

Nagaoka, in Japan, also saw both gold and something that looked like silver. The next chapter tells Nagaoka's story, and the following chapter concludes the Miethe story.

CHAPTER 21

Golden Journey

Nagaoka Finds Visible Specks of Gold While Investigating Mercury Isotopes (1924-1925)

The news of Adolf Miethe's transmutation of mercury to gold appeared in the *Morning Post* on July 21, 1924. *Naturwissenschaften* had published the article three days earlier. *Nature* reviewed the paper on Aug. 9, 1924, and in December 1924, *Scientific American* announced its intention of sponsoring a replication. Those are the key events described in Chapter 20.

But Miethe wasn't the first scientist in 1924 to suggest the possibility of a mercury-to-gold transmutation; physicist Hantaro Nagaoka (1865-1950), in Japan, was. Nagaoka mentioned the idea in the March 29, 1924, issue of *Nature* and published his preliminary paper a year later.

Nagaoka mentioned the transmutation concept as a potential outgrowth of other research he was doing, spectroscopically investigating isotopes of mercury. Nagaoka and his colleagues, while performing their experiments, noticed an anomaly in the spectrum that suggested the possibility of transmutations. At the end of their paper, the authors wrote, "It should be pointed out that, if the above assumption as to the mercury nucleus is valid, we can perhaps realize the dream of alchemists, by striking out a H-proton from the nucleus by alpha-rays, or by some other powerful methods of disruption." (Nagaoka, Sugiura, and Mishima, 1924)

Within two months, simply on the basis of that one sentence, Nagaoka attracted a challenger. Carl David Tolmé Runge, a German mathematician, physicist, and spectroscopist, wrote a letter to the editor of *Nature*, expressing his critique of Nagaoka's as-yet-untested transmutation hypothesis. (Runge, 1924) But Nagaoka hit back hard and accused Runge of playing games with numbers in order to sow seeds of doubt and uncertainty. (Nagaoka, 1924) It took a year for Nagaoka to perform and report the results of his experiments.

Glimmering Specks

Nagaoka was neither a crackpot nor a young lad fresh out of college. In 1925, at 60, he was known and respected worldwide and was finishing 24 years as a physics professor at Tokyo Imperial University. He was then appointed head scientist at Rikagaku Kenkyūjo (RIKEN) in Komagome, the preeminent research institute in Japan, also known as the Institute of Physical and Chemical Research. He also served as the first president of Osaka University from 1931 to 1934. A 1925 article in the *Montreal Gazette* referred to Nagaoka as the "Japanese Einstein."

In 1904, long before he ventured toward the outer bounds of science to explore transmutation, Nagaoka gained worldwide recognition for his attempt to develop a model of the atom, which was published in the British journal *Philosophical Magazine*. This gave him membership in an elite circle of great minds in physics. That same year, Joseph John Thomson also came up with an idea for an atomic model, and in 1911, Rutherford suggested a model, as well. Eventually, in 1913, Niels Bohr, a Danish physicist who had worked under the guidance of Thomson and was then working in Rutherford's laboratory, developed a viable model of the structure of the atom that is still used today. Bohr was awarded the Nobel Prize in physics in 1922 for this work.

According to a paper by Charles Baily when he was a post-doctoral researcher at the University of Colorado, Boulder, Nagaoka's modeling work played a useful role. "Although all of the models [of Thomson, Nagaoka, Rutherford and Bohr] were later shown to be incorrect or incomplete," Baily wrote, "each one represented an essential step toward an understanding of the nature of matter, a view of the physical world often taken for granted a century down the road." (Baily, 2008)

Hantaro Nagaoka

In the first of Nagaoka's papers, he explained that he and his colleagues, rather than attempting to prove the alchemist's dream, were merely seeking insight into the atomic structure. They were examining spectroscopy lines of mercury and bismuth. It must certainly have been quite a surprise to them when they saw spectral lines of gold. (Nagaoka, 1924)

Nagaoka's electric discharge apparatus that allegedly produced artificial gold

After Nagaoka and his first group of researchers saw the initial telltale spectral lines of gold in March 1924, he continued to work with a second team, and this time, they prepared experiments specifically to try to make gold. They applied a high electric field to mercury atoms in the presence of a hydrogen-bearing material: transformer oil. The oil provided an abundant supply of hydrogen atoms, and the tungsten electrodes provided a host metal.

They believed they succeeded on Sept. 15, 1924. Nagaoka had thought the transmutation was going downward from mercury to gold, but it's much more likely that the transmutation was going up from tungsten to gold. The book *Hacking the Atom: Explorations in Nuclear Research, Vol. 1* goes deeper into a likely theoretical explanation. From the perspective of going up the periodic table, Nagaoka went from tungsten up four steps to platinum and a fifth step to make stable gold. He published his group's results in *Nature* as a letter to the editor on July 18, 1925. Here's how he described it.

"The mercury gradually turned into fine globules until the oil and mercury were mixed into a black pasty mass," Nagaoka wrote. "Mr. Yasuda, an expert in gold assaying, showed us minute gold specks extracted from the black mass obtained in the experiment of the previous day." (Nagaoka, 1925)

They were able to see specks of gold, embedded in ruby glass, at 150 times magnification. This was far more than just a few atoms.

Nagaoka provided additional details about these images: "The white ring [in the first photo] is greenish blue, and the lightly shaded one is rosy; these colors are characteristic of gold colloids. Numerous spots of this kind are obtained in the bottom of the distilling flask during the aftertreatment of a mixed mass of carbon and mercury after heavy condensed discharges. Sometimes ruby glass is apparently covered with a thin film of gold; and on microscopic examination, it is found to consist of fine particles of gold very densely distributed."

Japanese postage stamp commemorating Nagaoka's Saturnian model of the atom

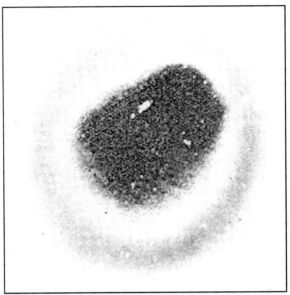

Central dark area is the spot of ruby glass photographed with transmitted light, magnified 150 times. The center of the spot contains specks of gold particles.

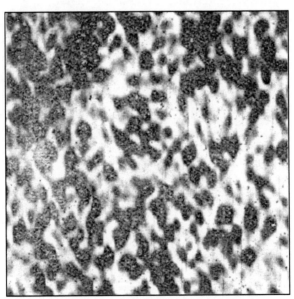

Portion of image, above, magnified at 2,500 times.

A Black Pasty Mass

Lewis Larsen, a theorist from Chicago, who reviewed part of this Nagaoka history, summarized the experiment in a 2012 slide presentation:

> In simplest terms, Nagaoka created a powerful electric-arc discharge between a spark gap comprising two metallic, thorium-oxide-free tungsten electrodes bathed in a dielectric liquid: either paraffin, kerosene or transformer oil that was "laced" with liquid mercury.
>
> Depending on the experiment, arcing between tungsten electrodes in oil was continued for 4 to 15 hours until "the oil and mercury were mixed into a black pasty mass." Note that mercury readily forms amalgams with many different metals, including gold and tungsten.
>
> Small flecks of gold were sometimes quite visible to the naked eye in "black masses" produced at the end of a given experiment. They also noted that "the gold obtained from mercury seems to be mostly adsorbed to carbon."
>
> Microscopic assays were conducted by "heating small pieces of glass with the carbon," to form a so-called "ruby glass" that can be used to infer the presence of gold colloids from visual cues very apparent under a microscope.
>
> Critics complained about the possibility that the gold observed was some sort of contamination. Responding to critics, Nagaoka and his group further purified everything they could think of and also made certain that the lab environs were squeaky clean; they still kept seeing anomalous gold. In some experiments they also observed "a minute quantity of white metal." Two years later, in 1926, Nagaoka reported to *Scientific American* that they had finally been able to identify the white metal — it was platinum. (Larsen, 2012)

Nobody at the time, including Nagaoka, had any definitive idea how these transmutations might have occurred. The idea that a proton could somehow overcome the Coulomb barrier at room temperature was as inexplicable and improbable in the 1920s as it is today. A viable idea about how neutrons could be created in such relatively mild conditions (and subsequently decay to protons) wasn't explained until Allan Widom and Lewis Larsen published their theory in 2006. (Widom and Larsen, 2006) Neutrons weren't even recognized until Chadwick discovered them in 1932.

Alternatively, high-energy physics could easily provide the necessary energy for a proton to overcome the Coulomb barrier at room temperature. In 1941, researchers at Harvard bombarded fast neutrons at mercury to transmute tiny bits of it into gold; however, the gold isotope was unstable. (Sherr, 1941) Larsen analyzed the significance of the Nagaoka experiment:

> Unlike the comparatively unknown Wendt and Irion team at the University of Chicago, Nagaoka was a world-renowned physicist and one of the most preeminent scientists in Japan. Nagaoka was a contemporary and direct competitor of Rutherford; Nagaoka's Saturn Model of the atom was the only competing model cited by Rutherford in his seminal 1911 paper on the nucleus.
>
> Given the very international character of science even at that time, it is very likely that Nagaoka was aware of worldwide controversy swirling around Wendt and Irion's exploding-wire experiments and of Rutherford's short but devastating critical attack on them in *Nature*. Lastly, Nagaoka must have known about Miethe and Stammreich's work in Germany. (Larsen, 2012)

Nagaoka closed his preliminary note with a humble yet far-sighted vision:

> The [high-current electric arc] experimental procedure here sketched cannot be looked upon as the only one for

effecting transmutation; probably different processes will be developed and finally lead to industrial enterprises. At present, there is no prospect of producing gold economically from mercury. Experiments with various elements may lead to different transmutations, which will be of significance to science and industry. Meagre as is the result, I wish to invite the attention of those interested in the subject so that they may repeat the experiment with more powerful means than are available in the Far East. (Nagaoka, 1925)

Gold in the News

Nagaoka's journal article was published on July 18, 1925. But Nagaoka apparently started telling the newspapers about his claim earlier. *Science Service* wrote one of the early news reports on April 29, 1925.

> The artificial production of gold from mercury by the application of strong electrical forces was announced recently by Prof. Hantaro Nagaoka of the Institute of physical and chemical research, Tokyo. ... In a statement made to *Science Service*, Nagaoka says:
>
> "Although the yield is small, and can only be examined with a microscope of low magnifying power, mercury is transmuted into gold. After a few more experiments as to the best laboratory arrangement, the method will be made public. So far as the experiment goes, gold comes out in colloidal state and forms beautiful ruby glass. On evaporating the treated mercury, gold is deposited on the bottom of the vacuum distilling apparatus in a very thin film, which, viewed with nearly normal reflected light, shows the characteristic yellow color, and seen with transmitted light, the complementary greenish tint. It stands the usual Cassius purple test for gold.
>
> "It was only in an extremely intense electric field that

the experiment succeeded. In order to be sure of transmutation, repeated purification of mercury by distilling in vacuum at temperatures below 200°C is essential. Finally, I wish to state that there are many Tokyo papers and journals giving mistaken notions of the experiment, so that translations of different statements are not in the least to be trusted." (*Science Service*, 1925, Japanese)

Gold on Tour

Larsen found a news story published two days later, on June 20, 1925, in the *Montreal Gazette*. Nagaoka was touring the United States and on his way to Europe, giving away samples of his artificial gold along the way as gifts.

> New York, June 19 — Bits of porcelain containing microscopic specks of gold which once was mercury were presented to Dr. George F. Kunz today by Dr. H. Nagaoka of Tokyo, the Japanese Einstein, who succeeded in producing artificial gold by disintegrating atoms of quicksilver in a powerful electric field. The Japanese scientist, who sails early tomorrow morning on the Olympic for Europe to attend the International Research Congress at Brussels, brought with him a number of gold-specked fragments from the porcelain flasks in which the mercury was treated. In more than 200 experiments, the change from mercury to gold has been confirmed, Dr. Nagaoka said. (*Montreal Gazette*, 1925)

In the same article, Nagaoka took a shot at Rutherford, a rare instance of any prominent scientist publicly expressing doubt or uncertainty about Rutherford's atomic disintegration news.

"Dr. Nagaoka," the *Gazette* wrote, "also expressed doubt that Rutherford had succeeded in effecting 'transmutation' or the disintegration of atoms by bombarding them with alpha particles. He said there was no way of confirming, by chemical analysis or otherwise, the supposed changes in atoms resulting from the bombardment by the alpha particles given off by radium."

It was a valid point, and Nagaoka was one of a few, if not the only scientist, to point out the obvious: Knocking particles off of an atomic nucleus, as Rutherford claimed he did (it's not exactly what happened; we'll get to that in Chapter 22), is not the same as changing the atom from one element to another. And in fact, Rutherford had no data about the changed residual nucleus.

> # SHOWS TRANSMUTED GOLD IN NEW YORK
>
> ### Dr. Nagaoka of Tokio Taking Specimens to Brussels Congress

News headlines from June 20, 1925, Montreal Gazette news story

As the *Gazette* article reported, Nagaoka went to Brussels to speak at the Congress of the International Union of Pure and Applied Physics. Nagaoka had been a member of the first executive steering committee of that organization back in 1922. He shared the company of other distinguished scientists on the committee such as Sir William Bragg, a British physicist, chemist, and mathematician who won the 1915 Nobel Prize in physics, Louis "Marcel" Brillouin, a French physicist and mathematician, and Robert Millikan, an American experimental physicist who won the Nobel Prize in physics in 1923.

After the general assembly at the congress, two papers were given: one by Hendrik Antoon Lorentz, a Dutch physicist and winner of the 1902 Nobel Prize in physics, and the other by Nagaoka. The meeting program could not have listed a more radical topic: "Transmutation of Mercury to Gold."

Golden Gift

George Frederick Kunz (1856-1932), the man to whom Nagaoka presented a specimen of his synthetic gold, was the research curator of gems and precious stones at the American Museum of Natural History in New York. Here's what *Journal of the Mineralogical Society of America* wrote in 1925:

> Dr. George F. Kunz has obtained what is reported to be the first sample of synthetic gold which has reached this country. It will form a part of the collection of elements at the American Museum of Natural History in New York. The sample of synthetic gold comes from the laboratory of Professor Hantaro Nagaoka of the Tokyo Imperial University. (*Journal of the Mineralogical Society of America*, 1925)

George Frederick Kunz

Kunz was well-known and respected in the mineralogical field for almost 60 years. At the young age of 23, he became a vice president of Tiffany and Company, and he delighted in allowing friends in his mineralogical circles to hold the 128-carat Tiffany yellow diamond (one of the largest diamonds in the world) while visiting the famous New York store.

According to the *Journal of the Mineralogical Society of America*, "Kunz was president of the association to introduce the metric system into United States, and all jewelers owe him a debt of gratitude for his efforts in aiding the establishment of the international carat." The gem kunzite was named in his honor, and he was instrumental in the naming of three other stones. He was prolific and wrote more than 300 articles on gems and minerals as well as several books.

Larsen did an extensive literature and news search, and as far as he could tell, neither Rutherford nor anyone else questioned or challenged Nagaoka's experimental results, as Larsen wrote in his slides:

> Nagaoka's results were probably correct. As far as we can tell, no one questioned their veracity. If the *Montreal Gazette* accurately reported all of Nagaoka's statements, Hantaro and his RIKEN colleagues detected production of gold metal in 200 experiments that were conducted between September 1924 and June 1925.
>
> Given Nagaoka's high stature internationally and longstanding reputation as a careful experimentalist, it would appear rather implausible that he and his RIKEN team would mistakenly identify some other element as gold. As the *Gazette* reported, Nagaoka went on a world lecture tour and handed-out actual samples of produced gold. Third parties would have had ample chances to analyze the samples. If gold was not found to be present, someone would assuredly have questioned his claims publicly. No one did. Rutherford, a competitor of Nagaoka's who had attacked Wendt and Irion with a highly critical article in *Nature* in 1922, was uncharacteristically silent about RIKEN's results. (Larsen, 2013)

Larsen didn't understand the sudden termination of transmutation research soon after Nagaoka's work. "It is puzzling why this seemingly fruitful line of inquiry appears to have died out worldwide sometime before Chadwick experimentally verified the neutron's existence in 1932," Larsen wrote. "Oddly, it does not appear that anyone else ever tried to exactly duplicate Nagaoka's experiments. However, there were well-publicized failures to replicate Miethe and Stammreich's experiments."

The Nagaoka timeline is marked by three key dates:

Key Dates of Nagaoka's Transmutation Activity

March 1924
Nagaoka suggests modern alchemy is possible in *Naturwissenschaften* paper.

June 1925
Nagaoka tours the world, giving out gold samples and lectures.

July 1925
Nagaoka publishes paper on the transmutation of mercury to gold in *Nature*.

Meanwhile, at the time that Nagaoka was on his world tour, new activity was taking place on the German transmutation front.

CHAPTER 22

The Death of Modern Alchemy

A High-Profile and Highly Publicized Failed Replication Puts an End to the Early Transmutation Era (1925-1927)

In June 1925, Japanese physicist Hantaro Nagaoka was touring the world, giving lectures and handing out samples of gold he had transmuted. The following month, *Nature* published his scientific paper. Also, the Adolph Miethe mercury-to-gold transmutation story resumed in Germany.

Industrial Opportunity

On May 7, 1925, Siemens and Halske Aktiengesellschaft, the massive German multinational engineering and electronics company, acquired the rights from Miethe and his collaborator Hans Stammreich and filed a patent on "Improvements in or Relating to the Extraction of Precious Metals." (Siemens, 1925)

A month later, on June 12, 1925, Siemens filed another patent, titled "A Process for Converting Mercury Into Another Element." This patent application did not list Miethe and Stammreich as inventors. (Siemens, 1925, June 12) Neither patent was issued.

News reports confirm that Siemens attempted experiments, and two references state that it succeeded in producing artificial gold. (Sheldon and Estey, 1926; *Nature*, 1926, 758-60, Present Position)

The first account of the Siemens' attempt comes from a wire service story, on July 14, 1925. Here is the full text of that news article:

Berlin, July 7 — Successful experiments in extracting gold from quicksilver on the basis of the recent discovery by professor Adolf Miethe are being conducted in the laboratories of the Siemens and Halske Works here under the direction of Dr. Kurlbaum, the German [physicist], it was learned today.

Dr. Kurlbaum bombards large surfaces of quicksilver with electrons, the amount of gold extracted depending on the strength of the bombardment. These experiments, which may lead to commercial exploitation of Miethe's discovery, have been made possible by Miethe's recent improvement on his own discovery, which he announced last month at a meeting of the Association of German Chemists.

[In June], Miethe said he had reduced the ratio of gold to quicksilver ... from one part in 10,000,000 to one part gold in 10,000, thus increasing the amount of gold obtained by 1,000. This discovery has now prompted Siemens and Halske to seek further development to the point at which it can perhaps become commercially profitable. (*Kingston Gleaner*, 1925)

Something to Worry About

Did Miethe really improve the efficiency of his gold-making process by three orders of magnitude? This would be a good time to recall the concern of the economists about the possible disastrous results of a sudden introduction of inexpensive artificial gold. They had said there was nothing to worry about unless the efficiency of the process could be scaled up dramatically. If Miethe did in fact increase the efficiency as reported, this is precisely the sort of thing that could have kept the world's financiers awake at night.

Later that July, four more transmutation papers published in the scientific journals. On July 17, 1925, *Naturwissenschaften* published a paper by Miethe as well as a paper by Otto Hönigschmid (1878-1945) and Eduard Zintl (1898-1941) at the University of Munich that discussed the atomic weight of Miethe's gold. (Miethe, 1925; Hönigschmid and Zintl, 1925) The other two papers were Nagaoka's preliminary reports, published on July 31 in *Naturwissenschaften* and on July 18 in *Nature*.

On Aug. 21, 1925, the Associated Press distributed a story that one news outlet titled "GERMAN SCIENTISTS FIND GOLD FROM MERCURY EASY":

> New York, Aug. 21 — (Associated Press) — Increasing success of German scientists in obtaining gold from mercury is described in dispatches to the American Chemical Society, made public today. Though far from being commercially profitable, the experiments of professors Adolf Miethe and Stammreich now are yielding 10,000 times as much gold from the same amount of mercury as they did a year ago. Such an eminent skeptic as professor Fritz Haber, whose synthetic ammonia process played a large part in German war plans, the society reports, has been won over.
>
> By the improved process, electrical discharges are sent between mercury electrodes in a dielectric, like paraffin. The gold is found in the mercury atomized in the spark path in the ratio of one part of gold to 10,000 parts of mercury. From 1 kilogram of mercury, one-tenth of a gram of gold can be obtained. Virtually all of the quicksilver is recovered and is used over and over again.
>
> The same results, the society reports, have been obtained at the Siemens Works, Berlin, when mercury surfaces were bombarded with electrons in extremely high vacuum.
>
> [After w]atching professor Nagaoka at Tokyo, who discovered the gold-producing process independently of German savants, professor Haber is quoted as saying that he has confirmed the results by repeating the experiment himself. Professor Nagaoka has detected a second substance, similar to platinum, in the gold residue. *(Gastonia Daily Gazette,* 1925)

Stamp of Approval

After the Associated Press story, the *New York Times*, on Aug. 22, 1925, wrote a more detailed account and reported that Haber, who had checked Miethe's gold a year earlier, had now checked and confirmed Nagaoka's experiment and results.

> **GOLDMAKERS SAY SUCCESS IS GROWING**
>
> Extraction of Precious Metal From Mercury Is Increased 10,000-Fold, Germans Report.
>
> **COST IS STILL TREMENDOUS**
>
> Dr. Sheldon Expected to Report Here in a Month on Results From a Baser Metal.

Headline from August 22, 1925, New York Times *news story*

The *Times* story also said that Miethe had recently reported his results before the German Chemical Society, according to a message from the American Chemical Society. The *Times* wrote that one of the societies (it was not clear which one) was going to publish the following message in its official journal:

> "The skepticism which was fostered privately and publicly toward the experiments of professors Miethe and Stammreich on the transformation of mercury into gold dwindled when investigators reported before the German Chemical Society on the results of their more recent experiments.
>
> "Professor Fritz Haber, internationally famous for his

development of synthetic ammonia, which played an important part in Germany's war plans, who previously cherished the greatest doubt as to the accuracy of the experiments, congratulated professor Miethe and related how, on his world tour, he had seen in the laboratory of professor Nagaoka at Tokyo, the apparatus with which the latter, but independently of Miethe, had likewise obtained gold from mercury, and that he himself could confirm the results by repetition of the experiment.

"The silver-like substance which often appears with the gold, or is formed almost exclusively, arises likewise, according to Miethe, from the mercury. Professor Nagaoka also has obtained, together with gold, a second substance which he described as similar to platinum." (*New York Times*, 1925, Gold Makers)

All About Miethe

The simultaneous independent finding of what appears to be platinum, in addition to gold, is a significant correlation. The finding also has bearing on an April 13, 1925, news report of a transmutation at the Kent laboratory at the University of Chicago. Researchers there found a transmuted metal with the weight of gold but the color of silver:

> Workers at the Kent Chemical Laboratory at the University are experimenting with a new machine. They already have been able, by its use, to change the weight of mercury put into the machine, but so far are baffled in their efforts to change the metal's appearance. The mercury now transformed, it is said, has the weight of gold though it does not look like it. (*Brandon Daily Sun*, 1925)

The next piece of news is a letter written by Miethe for the United Press and published, among other places, in the *San Mateo (California) Times*, on Aug. 26, 1925. Miethe began the letter by mentioning that he had been in contact with American scientists who were attempting replications. In his letter, Miethe asserted that Rutherford's disintegration from 1919 did not shift scientific consensus about the idea of an indivisible and immutable atom. "It was only following the successful disintegration of the atom by my experiments," Miethe wrote, "that the dogma of the indivisibility and unalterability of the elements collapsed."

Miethe's claim was quite self-centric but not unique; nearly every transmutationist in these early decades claimed that they were the first to accomplish modern alchemy. Many of the competing scientists — Nagaoka for example — often sowed doubts and uncertainty about their competitors' claimed achievements. Miethe would not be the last to make such a claim of priority in the world of transmutation research.

Miethe also explained that only after he and Stammreich carried out their experiments in the spring of 1924 did he hear about the similar claim from Nagaoka. It probably is so, for Miethe published his claim in a journal a year before Nagaoka did. Miethe was optimistic about the future. "In view of the situation," Miethe wrote, "many more scientists will now be encouraged to take up this matter."

Extraordinary Replication

A *Science Service* story published in, among other outlets, the *Berkeley Daily Gazette* on Sept. 15, 1925, tells of another related replication of the Miethe experiment. Arthur Smits and Albert Karssen at the University of Amsterdam followed the general method described by Miethe, but instead of starting with mercury, they started with lead. (Smits, 1925; Smits and Karssen, 1925; Smits 1926, Jan. 2; Smits 1926, May 1)

Based on the news report, Smits and Karssen measured the changing elements spectroscopically, *in situ*, in real time. "[After] six hours, lines characteristic of mercury began to appear in the spectrum," the paper reported. "These gradually strengthened until after 10 hours, the entire series of mercury lines and also those of the rare element thallium were perceived in the visible and ultraviolet parts of the spectrum."

Quartz-lead lamp used by Smits and Karssen to transmute lead to mercury

Mercury is two steps down the periodic table from lead; thallium is just one. Of all the known elements, the odds of two neighboring elements appearing during the experiment merely as "contamination" rather than transmutation are extraordinary. This was a strong confirmation of the Miethe-Stammreich method, which also transmuted a starting element (mercury) down one and two steps, to gold and platinum, respectively. Two independent experiments showing two similar sets of neighboring elemental changes place the odds of coincidence exceedingly high.

195.09	196.97	200.59	204.37	207.19
78	79	80	81	82
Pt	Au	Hg	Tl	Pb
Platinum	Gold	Mercury	Thallium	Lead

Section of the periodic table from platinum to lead

The Smits and Karssen replication was not fully appreciated at the time. The *Science Service* article also mentioned "a new sharp attack on

Miethe's experiments" from the laboratory of Ernst Hermann Riesenfeld, Wilhelm Haase, Ehrich Tiede, Arthur Schleede and Frieda Goldschmidt, at the University of Berlin Chemical Institute. They tried to replicate the experiment, but they failed to produce any gold. The researchers at the institute claimed that their measurements were more accurate than those of Smits and Karssen. The Chemical Institute group suggested that, because its own replication failed, Miethe and Stammreich's experiment was therefore a failure.

Nothing but a Dream

On Oct. 20, 1925, the Associated Press wrote a story, published in, among other outlets, the *Kansas City Star*. The experiments at New York University failed to replicate Miethe's results, according to an Oct. 19, 1925, press release from *Scientific American*. The magazine had rendered its verdict and, by the failure of its sponsored scientists to produce gold, deemed the original experiment by Miethe to be nothing but a fantasy, according to the *Star's* story headline "Nothing To Replace Gold":

> The age-old dream of the alchemist is still nothing but a dream, a statement issued yesterday by the *Scientific American* asserts. There is no fact now visible on the scientific horizon which threatens gold as a world standard of money. ... The experiments were conducted by professor H. H. Sheldon of New York University and Roger S. Estey, a [physicist], and others, the result being a "complete failure to confirm the transmutation of mercury into gold as announced by professor Miethe." (*Kansas City Star*, 1925)

In just a single day, despite encouraging reports of four replications — the University of Amsterdam, the Institute of Physical and Chemical Research in Japan, Siemens, and the Kent laboratory at the University of Chicago — the idea of gold-producing low-energy transmutations was dead. And all it took was one prominently reported failure, conducted at a reputable university, under the auspices of a prestigious science magazine. Here's the first paragraph from the *Times* story:

The hope that gold may be produced from mercury by any process known at present has been exploded as a result of exhaustive tests made by professor H.H. Sheldon, chairman of the Department of Physics at New York University, and Roger S. Estey, an instructor in the university, it was announced yesterday by the *Scientific American*, which supplied the funds for the tests. (*New York Times*, 1925, Gold From Mercury)

> **GOLD FROM MERCURY FOUND IMPOSSIBLE**
>
> New York University Tests Show That Transmutation Method Does Not Work.
>
> ---
>
> MIETHE PROCESS USED
>
> ---
>
> His Discovery of Gold Traces Laid to Use of Spanish Mercury, Which Contains Gold.

News headline from October 20, 1925, New York Times *article in response to* Scientific American *press release*

Failure to Communicate

More than a failure to replicate the experiment, this was a failure of science reporting. Sheldon and Estey did not identify any errors of protocol or analysis in Miethe and Stammreich's work or find that Miethe and Stammreich had overlooked a key assumption. Sheldon and Estey simply failed to replicate the results, and there are innumerable potential reasons why. Moreover, the *New York Times*, and perhaps other newspapers, was wrong to suggest that any failed experiment could render a hypothesis impossible. *That* is impossible.

But readers who look carefully at the *New York Times* story will argue that the physicists at New York University did say that Miethe and Stammreich overlooked a key assumption. "[Sheldon and Estey]," the *Times* wrote, "used the method described by professor Miethe in his published papers, but obtained no results and believed that the reason he found gold was because he used Spanish mercury, which contains a small amount of gold. The experimenters here used California mercury, which is free from gold."

Sheldon and Estey's explanation was flawed. Miethe and Stammreich and other scientists checked for gold in the mercury before and after experiments, as *Nature* reported: "An amount of the original mercury equal to that which was removed from the lamp was analyzed and found to be free from gold both by Miethe and independently by K. A. Hofmann." (*Nature*, 1924, 197-8)

Spreading the Word

Five days later, the *New York Times* reported that the group at University of Berlin Chemical Institute was again attempting to discredit the Miethe claim. The *Times* reported that even Nagaoka was going on the offensive, with a critique of Miethe. Nagaoka claimed that his experiments were more precise, and they probably were; nevertheless, his self-interest was evident. The *Times* quoted a statement from the American Chemical Society:

> "Ehrich Tiede, Arthur Schleede and Frieda Goldschmidt, on repeating Miethe's work, could in no case detect the formation of gold, and pronounce such formation, as in the accounts of Miethe and Stammreich, as at least difficult to reproduce. Likewise, professor E. H. Riesenfeld and his collaborator W. Haase, declare that, according to their experience, mercury is to be obtained practically free of gold only after many repeated slow vacuum distillations, and that all mercury preparations hitherto utilized for gold

production and described as free of gold contained that metal. Moreover, according to Riesenfeld, all mercury on the market contains gold." (*New York Times*, 1925, Discredit Gold)

Now the critics were contradicting one another: The Berlin chemists claimed that no mercury on the market was free of gold, yet the New York physicists had claimed they used mercury that was free of gold. Soon, another critic would claim that the gold actually came from the cathodes.

The November 1925 issue of *Scientific American* carried the full story of its sponsored failure to replicate. *Scientific American*, which called itself "the gold standard" for science reporting, reiterated its intentions: "Feeling itself obligated to discover the truth of this matter, both in the interest of science and of finance, the *Scientific American* arranged during the latter part of 1924 for a comprehensive and exact test of the results announced by professor Miethe."

DISCREDIT GOLD-MAKING BY MERCURY PROCESS

Headline from Oct. 25, 1925, New York Times *article*

After explaining how Sheldon and Estey did such a careful job with their experiment, *Scientific American* wrote at the end of the article: "It would be improper to assert on the basis of these results alone, that Professor Miethe's experiments have been proved to be definitely wrong. All that it is proper to say is that a careful, competent, and long, continued effort to confirm the German results has resulted in an entire failure to do so."

Of course *Scientific American* could not say that its sponsored failure could not prove that Miethe's experiment was "definitely wrong." More significantly, the failure revealed nothing about any error in Miethe's experiment. Replication failures, in and of themselves, can never reveal errors of an original experiment. Only analysis and inspection of an original experiment can reveal errors in the original experiment. This fundamental scientific concept was not (and is still not) broadly understood.

A few paragraphs later, *Scientific American* disregarded its very own advice about what was proper to say about the failure. "On the basis of all of the evidence now available," *Scientific American* wrote, "including the experiments of Dr. Sheldon and Mr. Estey which are here reported, it is our belief that a transmutation of mercury atoms into gold atoms does not occur and will not occur under the conditions which have been described by Professor Miethe."

What If?

In the November 1925 issue of Massachusetts Institute of Technology's *Tech Engineering News*, the MIT writer cautiously reported that *Scientific American*'s announcement made "it seem as though a slight amount of gold must have been present in solution in the original experiment in Germany." The MIT writer, who maintained an open mind, also realized the potential significance of Miethe's claim to the possibility of atomic energy.

"The most far-reaching effect of such an experiment," *Tech Engineering News* wrote, "should it prove successful, would not lie simply in the transmutation of mercury to gold, but rather in the opposite transmutation of gold to mercury, with the accompanying enormous energy liberated. Thus our fondest dream — atomic energy — would seem within reach." In Chapter 20, I wrote about the concept of transmutations going up the periodic table, in the context of Paul Darwin Foote. In fact, this idea comes up again a bit later in this chapter.

Making Mercury

In February 1926, *Scientific American* published another surprise. Despite the fact that it had written that transmutation of mercury to gold *was not* possible, it published an article suggesting that transmutation of gold to mercury *was* possible.

It printed a two-page article by Smits and Karssen, with data similar to that in their Aug. 7, 1925, *Naturwissenschaften* paper. The article gives no reason to doubt that Smits and Karssen had successfully transmuted lead into mercury and thallium.

Did the Smits and Karssen results support the Miethe claims? Yes, according to a Feb. 4, 1926, *Science Service* article:

> The Dutch experiments tend to support the claims of professor Adolf Miethe, of the University of Berlin, and professor Hantaro Nagaoka, of the Tokyo Imperial University, that they have changed lead into gold. When their claims were first made, many scientists doubted that such a change could be effected without the use of vast amounts of energy, far more than any of these modern alchemists had used. (*Science Service*, 1926)

About-Face

Also that month, Sheldon and Estey got more media attention for their failure. They reported it at the meeting of the American Physical Society on Feb. 27, 1926, and it got picked up by the Associated Press:

> Montreal, Feb. 27 (AP) — German methods claimed successful for the conversion of mercury to gold, are a flat failure, according to a paper prepared by Dr. H. Gordon Sheldon, and [Mr.] Roger S. Estey, of New York University, read before the American Physical Society here today. "During the past year we have repeated the mercury-to-gold transmutation experiments of Dr. Miethe of Berlin and we have not been able to produce the slightest trace of transmuted gold," they reported. Apparatus patterned after Dr. Miethe was used in their tests. (*San Jose Evening News*, 1926)

264 • THE DEATH OF MODERN ALCHEMY

Fritz Haber

On March 4, 1926, Fritz Haber (1868-1934), the independent German scientist who had endorsed the Miethe experiment, retracted his endorsement. Two years earlier, he had checked to make sure there was no gold contaminate before the Miethe experiment began, and he had confirmed the presence of produced gold after the Miethe experiment ended. Haber came up with an extraordinary explanation, according to the *New York Times*:

> Berlin, March 4 — That someone allowed his gold-rimmed glasses to touch the mercury with which professor Miethe performed his experiments of changing the atomic weight of mercury to produce gold is the opinion expressed

by Dr. Haber, the eminent scientist, after exhaustive tests. Dr. Haber believes there was a trace of gold in the mercury or that it came in contact with gold before or during the electrical charges and that their theory of being able to change the atoms as advanced by professor Miethe is totally incorrect. (*New York Times*, 1926)

The timing of Haber's retraction, just five days after Sheldon and Estey reported their failure at the American Physical Society, could be coincidence. On the other hand, perhaps Haber began to doubt himself and succumbed to peer pressure. He had little to gain and much to lose by maintaining his endorsement of Miethe and Stammreich's claim.

Reasonable Retrospective

Among the dizzying and often contradictory news reports about the 1924-1925 volley of transmutation experiments, one assessment written at the time stands alone and provides a sensible and objective view of the larger picture. On May 29, 1926, *Nature* wrote an article called "The Present Position of the Transmutation Controversy." (*Nature*, 1926, 758-60, Present Position) For readers who want to dive deeper into these events, this article in *Nature* provides an excellent roadmap.

The first section of the article summarized the experimental claims, spanning the work of Miethe and Stammreich, Nagaoka, Smits and Karssen, and a few others.

"Duhme and Lotz," *Nature* wrote, "are reported to have observed the formation of gold when a sufficiently powerful current is passed between the electrodes dipping into mercury; and investigators in the Siemens research laboratory have produced gold by bombarding a mercury surface with electrons in a very high vacuum."

Nature was forthright, and wisely so, in its preamble to its discussion of possible theoretical mechanisms. "A priori objections to these experimental results are not difficult to conceive," *Nature* wrote. *Nature* offered a thoughtful review of the experimental facts and possible theoretical explanations without prejudice.

Nature also analyzed the extant critique by the more prominent parties. In one such analysis, *Nature* discussed a reply by Miethe and Stammreich to the researchers at University of Berlin Chemical Institute. The Chemical Institute critics said they could detect gold in the starting mercury in their own experiments, and they implied that this invalidated Miethe and Stammreich's claim. However, "in their reply," *Nature* wrote, "Miethe and Stammreich ... point out that the microbalance which Riesefeld and Haase used for weighing their gold had an error of 0.003 mgm. and that this was exactly the weight of the gold [Riesefeld and Haase] extracted."

Spectacle Contamination

Nature devoted the last section of the article to a lecture given by Fritz Haber in Berlin and published on May 7, 1926, in *Naturwissenschaften*. Haber explained that one of his young collaborators, the only one in his lab who had found gold, had been contaminating his samples by touching his gold spectacles during the experiment.

Additionally, Haber speculated that the cathodes used in Miethe's experiments contained gold before the experiment began. To test this, Haber set up and replicated Miethe's experiment and found the presence of gold in the mercury afterward. *Nature* did not state whether Haber also tested for the presence of the silver-like material in the cathode, but it wrote that Haber believed it had a similar origin.

So now Haber had proffered two explanations for how gold appeared in his lab's experiments: It existed in one of the starting materials, and it was added by his bespectacled assistant.

It is certainly possible that gold preexisted in the cathodes of Miethe and Stammreich, Nagaoka and the researchers at Siemens. But Haber's proposal of occluded gold raises the question of why every experimenter who attempted a replication failed to find any gold.

Calling the Bluff

But there was a bigger problem with Haber's hypothesis of hidden gold in the electrodes. Milan Wayne Garrett, at the Clarendon laboratory of Exeter College, Oxford University, set up a test to see whether gold could diffuse from the electrode wires used in the experiments. Haber was widely recognized as a science authority by this time for his development of a process to produce nitrogen for fertilizer, thereby helping to avert a crisis in the food supply.

Garrett, on the other hand, was a doctoral candidate working under the direction of Frederick Alexander Lindemann (1886-1957), the director of the lab. Garrett's test of Haber's idea failed, and Garrett described, as diplomatically as he could, how his findings contradicted the world-famous Haber:

> Haber has also reached the conclusion that Miethe's gold came from his electrodes, and has given experimental evidence in support of his conclusions, though it must be admitted that this evidence is not altogether consistent with the results of the experiments described in [my] paper; for, if contamination from the materials of the seals occurs as readily as Haber has found, it is difficult to understand the uniformly negative results of [my] investigation. (Garrett, 1926, 404)

Haber had explained his alleged occlusion of gold in a 1926 paper in *Naturwissenschaften*. (Haber, 1926) Garrett had a lot of trouble believing Haber's explanation:

> The writer feels that some of the results of Haber stand themselves in need of further elucidation. In one experiment, in particular, [Haber] found an astonishing result. Here, 97 percent of the entire gold content of several grams of nickel and steel wire, employed in the seals of a hot-filament discharge tube, [somehow] diffused in some

extraordinary way to the surface of the wire, whence it evaporated and found its way quantitatively to the mercury anticathode, from which it was recovered by analysis. This surprising observation, if confirmed, would be capable of explaining in a perfectly satisfactory manner most, if not all, of the discordant results obtained by the various experimenters. ... Ordinary diffusion seems powerless to account for such a remarkable result, though it might possibly be brought about by some novel form of electrolysis. (Garrett, 1926, 404)

Garrett had great difficulty seeing any truth to Haber's hypothesis that the gold, in the quantities Haber measured, magically diffused to the surface, then migrated over to the anode. "It was proved," Garrett wrote, "by direct simultaneous determination that 10^{-8} gm of gold could have been detected, had it been present. No gold was found. Further results of Haber will be awaited with interest."

The verdict on the mercury-to-gold transmutation or, for that matter, the lead-to-mercury and lead-to-thallium, as well as the apparent creation of platinum, was far from final, as *Nature* wrote: "This short sketch of the main issues of the controversy will leave the impression that the question is still *sub judice* [under consideration/judgment]." (*Nature*, 1926, Present Position)

Reverse-Alchemy

The following month, something new happened. On June 27, 1926, the *Kansas City Star* reported a clever transmutation attempt with a twist. Instead of trying to make gold from mercury, Alois Gaschler, Miethe's student, tried to make mercury from gold and claimed to have succeeded. Instead of electron or alpha bombardment, Gaschler bombarded his target with protons, the *Star* reported:

> A reversal of the dream of the ancient alchemists, the transmutation of gold into a less valuable metal, mercury, is

claimed by Dr. A. Gaschler, an associate of professor Adolf Miethe, who, in 1924, announced that he'd succeeded in turning mercury into gold. Dr. Gaschler's process consisted in sealing a gold electrode into a vacuum tube and bombarding it with a stream of positive hydrogen atoms, shot through the tube at high velocity. The resulting color display, Dr. Gaschler watched through the spectroscope. At first the only light given off was of the color characteristic of glowing hydrogen, but at the end of 30 hours of bombardment, the spectrum lines that indicate the presence of mercury appeared and became stronger as time elapsed. (*Kansas City Star*, 1926)

In August that year, *Scientific American* picked up the story and published a long letter from Gaschler describing his work. I have not found any follow-up news stories or journal articles about Gaschler's work. Gaschler was granted U.S. patent 1,644,370 on Oct. 4, 1927, for a "method of artificially producing radioactive substances" with the use of a high electric current. He filed his application in 1924 and sought a German patent in 1923.

Finally, on Dec. 10, 1926, Sheldon and Estey, at New York University, published their paper claiming a failure to replicate the Miethe and Stammreich experiment. (Sheldon and Estey, 1926)

Scientific American, in a breach of scientific protocol, had issued a press release on behalf of Sheldon and Estey 13 months earlier. Miethe's opportunity to pursue further experiments came to a sudden end on May 5, 1927, when at 65, he died of heart failure immediately following surgery.

Critical Fissure

The chemists had made their best effort, and, with one exception that I will report in Chapter 24, this contentious research came to a close. The ownership of the field of radioactivity had been shared by chemists and physicists, but now things began to change. The inconclusive and doubtful spate of transmutation claims adversely affected the chemists' position in the field. Ownership of the field shifted to physicists, and, following Niels Bohr's insight into the atomic nucleus in 1921, the field of radioactivity later became known as the field of nuclear physics. The first paragraph of *Nature's* May 29, 1926, article "The Present Position of the Transmutation Controversy" captured the moment:

> If the genius of John Dalton gave the chemist a free-hold title to the atom, the work of Becquerel and the Curies may be said to have transferred the title to the physicist, or at least to have granted him an indeterminable [continuing] lease of the property.
> The physicist has made good use of his tenure: he has determined the structure and conditions of stability of the atom, and by embellishing its parts with attractive and repulsive signs, he has thrown light upon many things that were previously obscure, and revealed new avenues of research for the investigator. Except for some tentative efforts to formulate an electronic theory of valency, the chemist has, for the most part, been out of the picture; and even when
> Rutherford used alpha particles to disintegrate certain light atoms, the chemist was denied participation by the circumstance that the quantities of material involved were too minute to come within the range of his most delicate methods.
> In 1924, however, a vista of golden opportunity arose when Miethe and Stammreich announced that they had obtained from pure mercury gold in sufficient yield to be manipulated and determined by chemical means. The vista, although riddled by the barbed arrows of hostile criticism, is still above the horizon, and only the crucial test of further experimentation can decide its ultimate fate. (*Nature*, 1926, Present Position)

If the vista of modern alchemy was above the horizon in May 1926, within two years it had fallen below it. In March 1928, *Scientific American* wrote modern alchemy's short obituary, "THE RETREAT OF THE MODERN ALCHEMISTS."

But as far as I can tell, there was no mass retreat of the alchemists, with the exception of one 1927 incident that I will report on in Chapter 24. In the 1928 article, *Scientific American* reported an Oct. 1, 1927, retraction only from Smits, who, according to *Scientific American*, had "announced that his claim for successful transmutation was probably in error" because of contamination. *Scientific American*'s summary was narrow. Here is a more complete excerpt of Smits' summary:

> At the moment I am, therefore, inclined to conclude that the mercury found in our earlier sparking experiments came, certainly partly and perhaps entirely, from the carbon disulphide. ... But there is still this difficulty, that after sparking between the lead electrodes in carbon disulphide, purified in the usual way, the reaction was stronger positive than after sparking in this dielectric between platinum electrodes, in the same circumstances as regards voltage, current strength, and time. Consequently, there is still an uncertainty, which probably will be solved by our continued investigations. (Smits, 1927, 475)

Smits gave a complicated explanation of how he thought the mercury had come not from a transmutation of lead but from previously undetected traces of mercury in carbon disulphide which was used in the experiment. The excerpt above shows that he still could not explain why the results were stronger with lead electrodes than platinum electrodes.

I am not convinced that Smits' simplistic contamination scenario can sufficiently explain the fact that he and Karssen observed the gradually increasing appearance of new elements, *in situ*, during the course of the experiment, including the rare element thallium.

Scientific American ended its alchemy obituary with a sharp rebuke to members of the general public that might have been willing to consider the possibility of low-energy nuclear transmutations:

> This latest retreat from an advanced trench sets one aspiration of modern atomic physics in the rear of the alchemy of the middle-age philosophers. For they transmuted base metals into gold — or thought they did. In this connection, it may interest our readers to know that the alchemists are not all dead yet but, with the astrologers of 1928, are still going strong. Frequently the editor receives a little periodical wholly devoted to real alchemy. Despite our boasted methods of education, a fairly large proportion of the human race is still living, mentally at least, in the Tenth Century. (*Scientific American*, 1928)

With the exception of Smits and the 1927 incident, the modern alchemists did not retreat. Miethe was dead, and Nagaoka certainly never retracted, nor had any reason to retract, his claim. However, modern alchemy, as a field of scientific interest, did retreat. The 1924-1926 period of modern alchemy disappeared without resolution.

But the decline of modern alchemy did not occur in isolation; contributing factors were present. In the middle of the 1924-1926 wave of modern alchemy, a transmutation breakthrough of another sort was reported. This time, it was by a physicist, in 1925, at the prestigious Cavendish laboratory, under the direction of Rutherford. The transmutation happened not by the use of low-energy electrons but by high-energy alpha particles. And even though the Cavendish physicist observed a mere eight confirmatory particle tracks of a total of 400,000, his transmutation claim was accepted with relative ease.

The primary reason for the acceptance of the Cavendish transmutation was that alpha-induced transmutation was consistent with the understanding of nuclear theory of the day; electron-induced transmutations were not. A secondary reason for the acceptance, I believe, was that a mere handful of transmuted atoms would never threaten the economic stability of the developed world.

CHAPTER 23

Recovering a Milestone in Scientific History

Credit for the First Confirmed Transmutation Belongs Not to Rutherford but to His Student (1925)

As mentioned earlier, there is a widespread myth that in 1919 Sir Ernest Rutherford observed and reported the first man-made nuclear transmutation of a stable element. As the story goes, Rutherford bombarded alpha particles into a target of nitrogen and claimed that he observed the transmutation of nitrogen to oxygen, along with an ejected hydrogen particle, or proton. Some historians call this Rutherford's greatest achievement.

The Myth

He made no such observation and published no such claim. Here are some examples of the myth.

"In changing nitrogen atoms into oxygen atoms, Ernest Rutherford had become the world's first successful alchemist." (Campbell, 1999)

"Blackett started his research career in the 1920s with Rutherford, who had discovered the transmutation of nitrogen to oxygen." (Imperial College of London Web site retrieved March 26, 2014)

"The first artificial nuclear transmutation to be recognized and described as such was reported by Rutherford, 1919. ... The results of this experimental work correctly interpreted by Rutherford as a real nuclear disintegration of nitrogen nuclei into oxygen atoms and protons, resulting from the bombardment of nitrogen atoms by the alpha particles." (Sharma, 2001)

"Blackett, also at the Cavendish, used a Wilson chamber in the early 1920s to confirm Ernest Rutherford's transmutation of nitrogen into oxygen." (Westwick, 2005)

"In 1919, Rutherford succeeded [with] ... his wartime discovery that alpha particles from radioactive substances can transmute nitrogen into oxygen." (Heilbron, 2005)

"Blackett later proved, with the cloud chamber, that the nitrogen in this process was actually transformed into an oxygen isotope, so that Rutherford was the first to deliberately transmute one element into another." (Nobelprize.org, 2014)

The Fact

When Rutherford published his claim in 1919, he had no idea what happened to the nitrogen atom after the proton was emitted. His assumption that the alpha particle recoiled was wrong. He had no idea what mechanism caused the emission of the proton. His guesses about the nuclear mechanism were, in fact, wrong.

The discovery goes, instead, to Patrick Maynard Stewart Blackett (1897-1974), a research fellow working under Rutherford's direction. In 1925, Blackett, after performing experiments for four years and photographing several hundred thousand alpha particles, obtained the data and made the observation and claim of a nitrogen transmutation into oxygen. Blackett's data revealed that Rutherford's guess about the mechanism was wrong. Blackett published his results in 1925; Rutherford, who submitted Blackett's paper to the journal on his behalf, was not an author on the paper. (Blackett, 1925)

The scientific community gave Rutherford the credit for Blackett's transmutation discovery, and the myth took root. Below, I explain the facts and how I learned about this error in science history.

Assumption Versus Discovery

Rutherford had hypothesized that the bombarding alpha particle knocked a proton off of the target (a nitrogen atom), thus causing the alpha particle to recoil away. This would have produced a three-forked set of tracks. He did not, however, qualitatively test this. He made some guesses about the residual atom that were wrong.

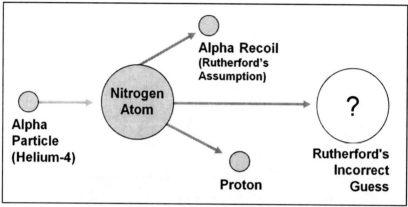

Conceptual diagram of the actual Rutherford observation and claim.

Blackett, however, performed and reported the tests that revealed a two-forked set of tracks, confirming that the nitrogen atom captured the alpha particle, temporarily transmuting nitrogen into fluorine, then, nearly instantly, ejecting a proton and leaving a residual oxygen atom.

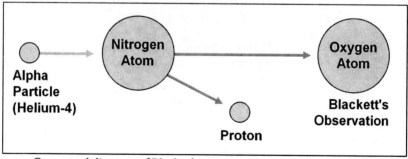

Conceptual diagram of Blackett's nitrogen-to-oxygen transmutation observation

All Rutherford observed, all he knew, was that a proton was emitted from the reaction process. Some people argue that Rutherford did artificially transform nitrogen to oxygen, even though he didn't know it at the time. Nevertheless, it seems incorrect to retroactively attribute any credit to him for something he didn't hypothesize, didn't observe, didn't claim, and didn't publish.

The Story

Initially, I had no idea that there was any myth. I was assembling the chronology of the key events in early nuclear transmutation history. (See Appendix E: Timeline of Key Events in Early Nuclear History) I was trying to find out when the first man-made nuclear transmutation of a stable element took place and who did it. Most print and online references incorrectly list the transmutation accomplishment as Rutherford's.

Comic strip frame from Atomic Energy Commission-sponsored publication

The myth even appeared in a U.S. government-sponsored comic strip about the developments of atomic energy. I went looking for the seminal paper to provide a reference for my chronology. My search for what I thought was going to be a simple scientific reference led me on a surprising journey. I had assumed that it would be easy to locate the seminal paper for such a landmark discovery. Far from it.

On the Hunt

Yet none of these accounts cited a reference for the transmutation discovery. I was puzzled: How could there be a major discovery like this without someone citing the reference? I contacted several science professors, including John Campbell, a Rutherford enthusiast, professor of physics at the University of Canterbury, New Zealand, and author of the book *Rutherford: Scientist Supreme*. Nobody seemed to know the scientific reference in which Rutherford published this discovery.

So I began to dig. I began with the scientific literature and found nothing. Then I searched the news archives and found some related events, but no mention of the announcement of the discovery. I took what I found from the news archives and went back to the scientific literature to examine those events.

Eventually, I found a paper by Rutherford, "Collision of Alpha Particles With Light Atoms. IV. An Anomalous Effect in Nitrogen," published in *Philosophical Magazine* in June 1919. From the title, I expected the anomalous effect to be the transmutation of nitrogen to oxygen because that was what all the excitement was about. To my surprise, there was no mention of a nitrogen-to-oxygen transmutation in the paper. The paper was the fourth in a series, so to be sure, I looked through all four for some sign of Rutherford's report of the oxygen transmutation data and his claim. I found nothing.

I went back again to the news archives. I found a reference in an August 1919 article in *The Independent*. The story mentioned the unconfirmed 1907 claims of Sir William Ramsay, as I discussed in Chapter 10. The news story noted that Marie Curie failed in her attempted replication of Ramsay's work and insinuated that her failure disproved Ramsay's claim. The article also suggested that Ramsay was sloppy and that he dropped cigarette ashes into his crucible.

The Hype

On the other hand, *The Independent* wrote confidently about Rutherford. The article gave me the first piece of the puzzle. Rutherford proved that a hydrogen atom (a proton) was knocked off a target nucleus when hit by a colliding alpha particle. But the article mentioned nothing about transmuting nitrogen to oxygen. After the August 1919 article in *The Independent*, I found no other news stories about Rutherford's June 1919 journal articles in the newspaper archives until I got to December.

On December 8, 1919, the Paris newspaper *Le Matin* published a Page 1, Column 1 story about Rutherford; apparently, the paper thought it had big news. (*Matin*, 1919) The article was written by Charles Nordmann. (Débarbat, 2007)

The article said nothing about any recent news that might have triggered the story. My search revealed no related news at the time or for the previous five months. After I obtained a translation of the article, I learned that the underlying story in *Le Matin* was the same as that reported in *The Independent*. They were both about Rutherford's shattering of a nitrogen nucleus and the emission of a proton.

Headline from Dec. 8, 1919, article in Le Matin: *"AN IMMENSE DISCOVERY - The Transmutation of Elements Has Just Been Achieved for the First Time"*

Le Matin took a different approach from *The Independent*. Whereas *The Independent* had simply and accurately reported the nature of Rutherford's claim, *Le Matin* spun the story as a modern-day discovery of alchemy and transmutation. The second sentence in the *Le Matin* story presented Nordmann's news angle.

"The transmutation of chemical elements," Nordmann wrote, "that old dream that preyed on the mystic minds of alchemists during the

Middle Ages, has just been accomplished for the first time. The media, those modern, telegraphic trumpeters of goodwill, have neglected to spread the news."

As far as I could tell, there was no triggering news event. Nordmann thought this was big news, and he blamed the lack of news coverage of Rutherford's discovery on his colleagues' "neglect." Yet in his article, too, there was nothing about transmuting nitrogen to oxygen.

I double-checked with Campbell, a Rutherford expert, to see whether he knew of any scientific publication by Rutherford on that day or during that period about the alleged transmutation of nitrogen into oxygen. He didn't know of any.

Nevertheless, the *Le Matin* story had a significant impact. Follow-up stories — all based on the *Le Matin* story — ran the next day in the U.S. from coast to coast. However, the *Le Matin* story had no news to report: no recent event, no science conference, no interviews, not even any direct quotes from sources.

Disintegration, Not Transmutation

Thus far in my investigation, I had found no evidence that Rutherford had transmuted one element to another. Eventually, I began to learn what Rutherford had actually done at the Cavendish laboratory in Cambridge. Historian Alfred Walter Stewart was very helpful. He described the novelty and importance of Rutherford's discovery: A proton was ejected as a result of the collision of the alpha particle and the nitrogen nucleus.

Someone Named Blackett

I was mystified. Where were the data, paper and claim of the transmutation of nitrogen to oxygen? The journals gave me no clue; the news archives didn't help, either. I began a wider search to see what other historians had written. The next historian whose work I found was Peter Galison. (Galison, 1997)

Galison wrote about Patrick Maynard Stewart Blackett. Blackett was born in London, England, on Nov. 18, 1897. He began his career as a naval cadet in 1914 and took part in World War I naval battles. Blackett graduated from Osborne Naval College in 1917, studied physics at Cambridge University, and earned his bachelor's degree there in 1921. After graduating, he worked as a research fellow in Rutherford's laboratory at Cambridge. (L'Annunziata, 2007, 359). Galison described an apparently crucial limitation of Rutherford's experiment:

> Rutherford made his discovery using the scintillation method, but evanescent flashes of light told him little about what was happening during the actual disintegration process.
>
> Blackett later remarked, "What was more natural than for Rutherford to look to the Wilson cloud method to reveal the finer details of this newly discovered process." [Takeo] Shimizu, a Japanese physicist working at the Cavendish [laboratory], modified the chamber so that he could quicken the expansions and take many more photographs. Shimizu also arranged the cameras to extract stereoscopic pictures from the cloud trails. Before he could do any more, however, he had to return to Japan, and the project fell to Blackett.
>
> Blackett raved about his new work: "Provided by Rutherford with so fine a problem, by Wilson with so powerful a method, and by nature with a liking for mechanical gadgets, I fell with a will to the problem of photographing some half-million alpha-ray tracks."
>
> The problem was to find a few golden events; most collisions between alpha particles and nitrogen nuclei would be elastic, but a few would not, and these would show Blackett and Rutherford what was happening in the supposed disintegration.
>
> After photographing 400,000 tracks and carefully examining them all, Blackett identified eight anomalous tracks that did not correspond to the elastic collisions. These were evidence of the transmutation of nitrogen [to oxygen]. (Galison, 1997, 118-9)

Then, to my surprise, Galison wrote that Rutherford's speculation about the process was wrong.

> Blackett discovered that the process was not one of *disintegration* but one of *integration*; only two tracks were seen after the interaction occurred. [This meant] that the alpha particle was absorbed as the proton was ejected. The resulting nucleus was a heavy isotope of oxygen.
>
> All Rutherford could know from his scintillation experiments was that the alpha particles infrequently caused nitrogen nuclei to emit protons — he could not see the actual interaction — he had assumed it was a disintegration process.
>
> Only the cloud chamber could provide a visual representation of the transmutation process itself and give physicists the chance to discover the intricacies of the exchange. (Galison, 1997, 119)

Patrick Maynard Stewart Blackett

Was it True?

I was confused. If Galison was correct, then the history of the first confirmed nuclear transmutation of a stable element, the 1912 and the 1924-1926 transmutations notwithstanding, was wrong, on three counts: 1) Rutherford's hypothesis and claim of a disintegration process was actually an integration process, 2) Blackett, not Rutherford, made the observation and discovery of the transmutation, 3) Blackett, not Rutherford, published the discovery of the transmutation. If true, these were serious contradictions. I began an even wider search, using the term "Blackett."

My next reference was the 2008 *Biographical Encyclopedia of Scientists*. Here's what it said:

> Just as Blackett was beginning his research career, Rutherford had announced his discovery of the atomic transmutation of nitrogen into oxygen by bombardment with alpha particles. Blackett, using a cloud chamber, took some 23,000 photographs containing some 400,000 alpha particle tracks in nitrogen and found, in 1925, just eight branched tracks in which the ejected proton was clearly separated from the newly formed oxygen isotope. (Daintith, 2008, 73)

The second part of the Daintith reference, the description of the photographs and the eight anomalous tracks, was consistent with Galison. However, Daintith said that Rutherford "announced his discovery" of the transmutation. But when? Where? How? I had found no such announcement from Rutherford — not in the science journals, not in the news media. I had more questions than answers. How could Rutherford have announced the discovery in 1919 if Blackett didn't find the tracks until 1925? And did Blackett publish in 1925 or some other year? Did he publish on his own, did Rutherford publish on his own, or were they co-authors?

Clearly, there were some serious inconsistencies, and I was certainly missing pieces of the puzzle. I needed a specialist. I needed someone who was an expert in the nuclear research performed in the first few decades of the 20th century. I wanted someone who, ideally, had been alive at the time, as well. I also needed a historian who was independent of the principal parties, so my search excluded biographical and autobiographical accounts of the key players.

Help From the Dead

I found the person I was looking for. More precisely, I found his book; he had died in 2005. Milorad Mladjenovic was born in Sarajevo (in present-day Bosnia and Herzegovina) in 1920, the year after Rutherford's big discovery. In 1992, Mladjenovic wrote *The History of Early Nuclear Physics, 1896-1931*, published by World Scientific. Of more than a dozen historical references I have reviewed from this period, nobody has chronicled this period of history more objectively, completely and accurately than Mladjenovic.

In his book, Mladjenovic tells the story of Rutherford's discovery, beginning with the experiments of Ernest Marsden, who had observed scintillations from hydrogen atoms in 1915, before Rutherford began studying the subject. But Marsden did not conclusively eliminate a key alternative explanation, as Mladjenovic explained:

> Rutherford published in 1919 a paper titled "Collision of Alpha Particles with Light Atoms," which shows step by step how he was led to the first artificial [disintegration]. The very beginning was due to some interesting results obtained by his former collaborator Marsden.
>
> In his first paper, Marsden reports the results of measurements performed with radium emanation enclosed in a thin glass tube; the movable emanation source was placed in a wide tube filled with hydrogen at variable pressure. A zinc sulfide screen was placed at the end of the tube, far beyond the extreme range of the alpha particles themselves in the hydrogen. The scintillations were observed, few in number and less

intense than those produced by alpha particles. In air [rather than pure hydrogen], no such scintillations were observed. The absorption measurements [showed] that the less-intense scintillations were produced by hydrogen atoms, or H-particles. (Mladjenovic, 1992, 157)

Rutherford was keenly aware of the alternate explanation that could account for the scintillations, and he proceeded meticulously to sort this out. He began his own experiments using helium and found no long-range recoil atoms. Then he hit pay dirt. He performed experiments with nitrogen and, separately, with oxygen and found bright scintillations clearly beyond the possible range of the alpha particle. Rutherford continued step by step, and he precisely and definitively identified protons being emitted, though in his paper he called them "hydrogen atoms." At the end of his June 1919 paper, Rutherford cautiously announced his discovery:

It is difficult to avoid the conclusion that these long-range atoms [protons] arising from the collision of alpha particles with nitrogen are not nitrogen atoms, but probably charged atoms of hydrogen or atoms of mass two. If this be the case, we must conclude that the nitrogen atom is disintegrated under the intense forces developed in a close collision with swift alpha particles, and that the atom [proton] liberated formed the constituent part of the nitrogen nucleus. (Rutherford, 1919, 586, Collisions IV)

No Interest in Transmutation

In 1921, Rutherford continued with his deep interest in the constituents of matter, particle physics and disintegrations. He began experiments with James Chadwick and looked at electrical charges, vectors, ranges and velocities of the emitted particles. But the two researchers were not looking for and did not find evidence of transmutation. They accomplished a lot, but that crucial piece was missing, as Mladjenovic wrote.

The description of transmutation itself was not complete, as it was not known what happened to the initial alpha particle after ejecting the [proton]. In their third paper, [Rutherford and Chadwick] state the following: "The fate of the alpha particle is a matter about which we have no information. It is unlikely that the field of force remains central at the very close distances. It is possible that the alpha particle is in some way attached to the residual nucleus. Certainly it cannot be re-emitted with any considerable energy, or we should be able to observe it." This problem was solved by Blackett. (Mladjenovic, 1992, 167-168)

Blackett Goes to Work

Rutherford directed Blackett to use the Wilson cloud chamber as the detection tool, as Mladjenovic explained:

> Rutherford and Chadwick examined motions involved in a collision, but they could not get direct information about them. The cloud chamber offered that essential missing information in a beautiful way; the way to the beautiful, however, was not easy. Since the probability of the transmutation of nitrogen was very small (about 20 per 1 million alpha particles), it was necessary to photograph a very large number of tracks. (Mladjenovic, 1992, 168)

Some years later, Rutherford wrote a concise description of the Wilson cloud chamber.

> The most wonderful of these methods has been devised by professor [Charles Thomson Rees] Wilson and depends on the observation that the ions produced by a fast particle act as nuclei for the condensation of water vapor upon them under certain conditions. Each ion then becomes the center of a visible drop of water.

Since a fast alpha particle produces more than 100,000 pairs of positive and negative ions in a gas, the actual track of the flying particle through the gas becomes visible as a crowded trail of water drops. (Rutherford, 1937, 20)

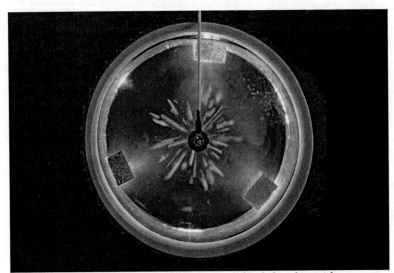

Looking down into a Wilson-type cloud chamber with radioactive source in the center. Courtesy Ron Galli

In their March 29, 1924, paper in *Nature*, Rutherford and Chadwick said nothing about any transmutation from nitrogen to oxygen. (Rutherford, 1924)

Mystery Solved

I soon found the Blackett paper. Mystery solved: He was the one who had observed and, on Feb. 2, 1925, published the discovery. It was a detailed, precise explanation of the experiment and results, and it included the now-famous photographs that Blackett obtained from the Wilson cloud chamber. Neither Rutherford nor Chadwick was a co-author on the paper. (Blackett, 1925) Galison — and a few other historians — were correct:

1) Rutherford's hypothesis and claim of a disintegration process, as the primary process, were wrong;
2) Blackett, not Rutherford, made the discovery of the transmutation;
3) Blackett, not Rutherford, published the discovery of the transmutation.

Blackett's smoking gun was the two-forked trails he saw in eight of the 400,000 tracks he photographed in the cloud chamber. In the paper, Blackett went into great detail to analyze the particle kinetics which further reinforced his analysis. He correctly observed, analyzed and published his discovery:

> In ejecting a proton from a nitrogen nucleus, the alpha particle is therefore itself bound to the nitrogen nucleus. The resulting new nucleus must have then mass 17 and, provided no electrons are gained or lost in the process, an atomic number of eight. ... It ought, therefore, to be an isotope of oxygen. (Blackett, 1925, 351)

Clearer Picture

In his paper, Blackett did not get so far as to write out a precise chemical equation for the reaction. Some historians have written the reaction as: $^{14}_{7}N + ^{4}_{2}He \rightarrow ^{17}_{8}O + ^{1}_{1}H$. But this is slightly over-simplified. This is the full reaction: $^{14}_{7}N + ^{4}_{2}He \rightarrow ^{18}_{9}F \rightarrow ^{17}_{8}O + ^{1}_{1}H$.

The nitrogen nucleus captures the helium-4 nucleus (alpha particle) and for a brief moment, becomes a fluorine-like nucleus, according to Rutherford. That nucleus is very unstable and nearly immediately spits out a proton, leaving a stable oxygen nucleus. (Rutherford, 1937, 34)

I colored Blackett's photo to illustrate the particle physics more clearly. An alpha (green line) approaches and stops when it collides with the nitrogen nucleus. The alpha (2 protons) is captured by the nitrogen nucleus (7 protons), momentarily creating a fluorine nucleus (9 protons).

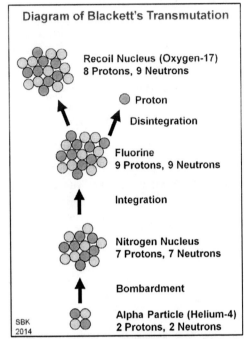

Diagram of Blackett's nitrogen-to-oxygen transmutation (Krivit)

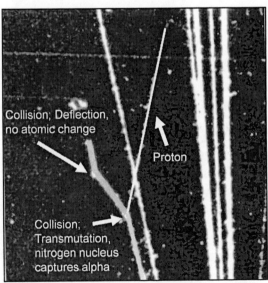

Photo by Blackett of an alpha-particle colliding with a nitrogen nucleus. The resulting oxygen nucleus goes left, and the proton goes right. (Color added)

A proton (yellow line, 1 proton) ejects toward the right, away from the fluorine nucleus, and the residual oxygen nucleus (blue line, 8 protons) recoils toward the left. The oxygen nucleus hits another nucleus but does not interact with it, and it simply changes course slightly until it comes to a stop. The key part of the discovery is that the fork is only two-pronged, not three-pronged. This is what Blackett saw that told him the alpha particle had caused the transmutation from nitrogen to oxygen.

Rutherford had thought that the incoming alpha particle would knock a proton off the nitrogen and all three particles would go their separate ways. That's why he called it a disintegration process. It's true that, when the proton ejected from the fluorine, a disintegration process took place, but that was a secondary reaction. The primary reaction was the capture of the alpha by the nitrogen, momentarily transmuting it to fluorine and, on the ejection of a proton, transmuting to oxygen.

Turning Point in Physics

Mladjenovic summarized the significance of *Blackett's* discovery:

> The discovery of artificial transmutation changed the nature of the activity of nuclear physicists; instead of observing which nuclear transformations were taking place in nature, physicists were [now] able to produce them themselves. The natural radioactivity was a small, limited field, but the artificial transmutation turned out to be virtually unlimited and remains alive, seven decades after the birth. It is one of those discoveries, the importance of which cannot be estimated; they are beyond the grading and pricing, so important in our lives. (Mladjenovic, 1992, 169)

In 1913, William Ramsay, John Norman Collie and Hubert Patterson had preceded Blackett in reporting man-made nuclear transmutations. Had Thomson not invented the occlusion hypothesis, he too would have shared that credit. But Thomson, a Nobel Prize winner in physics, and one of the most prominent physicists in the world at the time, dismissed all the 1913 claims and convinced most of the scientific community that the chemists goofed.

On the other hand, Blackett's 1925 discovery of the first artificial transmutation of a stable atom was accepted easily by the scientific community. But to my knowledge, Blackett's claim was broadly accepted before there was any independent replication. The claim was accepted on the basis of only eight of 400,000 tracks recorded over four years; it was hardly repeatable. Why was his claim accepted?

I propose several explanations. First, and foremost, the result conformed to the current nuclear theory: Only high-energy alpha particles offered the required energy to effect transmutations. According to theory, electron-based transmutations were inexplicable. Second, Cavendish was a physics laboratory. It had a credibility advantage over chemistry, which still suffered from the stigma associated with alchemy. Third, instead of using a spectroscope, a reliable detector used by chemists, Blackett used the cloud chamber, an instrument preferred by physicists. Fourth, Rutherford had already established credibility for the alpha bombardment process as a way to disintegrate an atom. Last, Blackett benefited from the influence of Rutherford and Thomson, both world-famous physicists and Nobel Prize winners.

And certainly, Blackett needed this clout because his claim of creating oxygen was in a far different class from Rutherford's claim of knocking off a proton. Rutherford and Soddy had demonstrated the natural disintegration of unstable elements in 1902. Ramsay and Soddy had demonstrated the natural nuclear transmutation of unstable elements in 1903. The claims of these chemists' transmutations had been front-page news for years. In contrast, was Rutherford's man-made disintegration of a stable element revolutionary? I cannot say it any better than historian Thaddeus J Trenn: "By 1919, such evidence was no more unexpected than the rising sun at dawn."

Sour Grapes?

Blackett's discovery was far more significant than proving that atoms were composed of constituent protons. Blackett's discovery forced the scientific world to accept the fact that man now had the power to create elemental transmutations. How did Rutherford respond?

On April 4, 1925, two months after Blackett's paper published, Rutherford wrote a letter to *Nature* about Blackett's claim of the nitrogen-to-oxygen transmutation. "It is not my intention here," Rutherford wrote, "to discuss the bearing of these results on the mechanism of disintegration, but rather to direct attention to other results and suggestions in connection with this important problem." (Rutherford, 1925, 493-4, Disintegration)

Rutherford seemed uninspired by Blackett's discovery and mentioned two previous reports by other researchers who had made cloud chamber photographs and observed only three-fork tracks. Rutherford attempted a comparison of Blackett's work to that of the other researchers:

> It is, of course, difficult to reconcile these photographs with the eight obtained by Blackett in which no third branch has been noted; but it may prove significant that the collisions photographed by [the other researchers] appear to have occurred when the alpha particle has lost a good deal of its range. It is obvious that there is still much work to be done to clear up these difficulties. (Rutherford, 1925, 493-4, Disintegration)

For Rutherford, Blackett's work was not a discovery but a "difficulty" and an "important problem." Rutherford's hesitance was one step short of Oliver Lodge's comment about the 1913 transmutations: "a great deal of proof will be required before it can be accepted."

Courage and Risk

By 1925, Rutherford was recognized as one of the greatest physicists to walk the planet. Blackett, four years out of school, was a newcomer to science. However, Blackett had the courage to go where Rutherford did not.

Rutherford's marginalization in *Nature* of Blackett's discovery was guaranteed to minimize the public's and scientific community's perception of Blackett's accomplishment. Four years of work and nearly half a million tracks were not enough for Rutherford:

> We must await the results of further detailed experiments to see how far such observations of scattering throw definite light on the problem of the mechanism of the disintegration collision. It seems clear, however, that a large amount of careful quantitative work as well as a great number of photographs of alpha-ray tracks will be required before we can hope to obtain such detailed evidence of the mechanism of such collisions and of the fate of the bombarding alpha particle for all the "active" elements. (Rutherford, 1925, 493-4, Disintegration)

Rutherford knew that Blackett's evidence had contradicted Rutherford's hypothesis that the residual atoms would be either boron or carbon. Not a minor embarrassment! Rutherford had sent Blackett's paper on Dec. 17, 1924, on his behalf to the *Journal of the Chemical Society Transactions.* On April 5, 1925, in an abstract for his lecture to the Royal Institution of Great Britain, Rutherford acknowledged that Blackett had figured out that the primary reaction was an integration rather than a disintegration:

> Since the proof that protons can be expelled from the nuclei of many light elements, the fate of the bombarding alpha particle after the disintegration has been a matter of conjecture. To throw light on this question, Blackett has

recently photographed the tracks of more than 400,000 alpha particles in nitrogen. ... In these photographs, the fine track of the proton was clearly visible, and also that of the recoiling nucleus, but in no case was there any sign of a third branch due to the escaping alpha particle. He concluded that the alpha particle was captured in a collision which led to the ejection of a proton. ...

These experiments suggest ... that the alpha particle is captured by the nucleus. If no electron is expelled, the resulting nucleus should have a mass $14 + 4 - 1 = 17$, and a nuclear charge $7 + 2 - 1 = 8$ — that is, it should be an isotope of oxygen. It thus appears that the nucleus may increase rather than diminish in mass as the result of collisions in which the proton is expelled. (Rutherford, 1925, 76, Studies)

"The work of Blackett," Mladjenovic wrote, "thus completed the picture of the whole process of transmutations. This was the first case of synthesis, or building up, of a heavier atom from a lighter one. A new, very important dimension of nuclear dynamics was opened, and Rutherford announced it with only one sentence." (Mladjenovic, 1992, 168)

Deliberately Vague

Even in 1937, in a small book, *The Newer Alchemy*, based on one of Rutherford's lectures, Rutherford made it clear that his discovery in 1919 was only that nitrogen emitted a proton when bombarded with an alpha particle, and that this constituted proof that protons were a fundamental component of atomic nuclei. He was deliberately vague in his 1937 book, however, about who discovered the nitrogen-to-oxygen transmutation.

> The atoms of several light elements were bombarded by a very large number of alpha particles. Under these conditions, I found evidence in 1919 that a few of the atoms

> of nitrogen were disintegrated with the emission of a fast hydrogen nucleus, now known as the proton.
>
> In the light of later results, the general mechanism of this transformation is well understood. Occasionally, the alpha particle actually enters the nitrogen nucleus and forms momentarily a new fluorine-like nucleus of mass 18 and charge 9. This nucleus, which does not exist in nature, is very unstable and breaks up at once, hurling out a proton and leaving behind a stable oxygen nucleus of mass 17. ... By photographing the tracks of several hundred thousand alpha particles in an expansion chamber filled with nitrogen, Blackett observed several clear cases of transformation of the nitrogen nucleus. (Rutherford, 1937, 33)

Rutherford went very quickly from "I found evidence" to "in the light of later results" without stating that it was Blackett who not only obtained these later results but also had the guts to state that he had observed a nuclear transmutation and published the data to prove it.

In the text quoted above, Rutherford credited Blackett for only the photography. This was more than a careless omission. A person who read only Rutherford's 1937 book would certainly think that the credit for the nitrogen-to-oxygen transmutation belonged to Rutherford.

Almost certainly, Rutherford's 1917-1919 experiments did, in fact, transmute nitrogen, via fluorine, to oxygen. But Rutherford had no such knowledge. He made no such observation, published no such data, and made no such claim.

And So It Was

And so history was written, for example, by Dahl in his book *From Nuclear Transmutation to Nuclear Fission, 1932-1939*:

> The guns of World War I fell silent in November of 1918, on the eleventh hour of the eleventh day of the eleventh month of the year. The Treaty of Versailles was signed seven months later, in June 1919. That June, marking

the formal end of the Great War, also marked the completion of a series of experiments in Manchester, England, by Ernest Rutherford. The publication of his epoch-making results that very month, also signaled the last piece of scientific work that Rutherford undertook at Manchester University. (Dahl, 2002, 5)

Dahl wrote about Rutherford's fascination with the possibilities of disintegrating atoms. Dahl mentioned nothing about any claim by Rutherford of a transmutation or creating oxygen from nitrogen. Here's how Dahl skirted the facts:

> The actual equation for the reaction, $^{14}_{7}N + ^{4}_{2}He \rightarrow ^{17}_{8}O + ^{1}_{1}H$, was only confirmed by P.M.S. Blackett, a junior colleague of Rutherford, in 1925. (Dahl, 2002, 9)

This is revisionist history. The equation was not *confirmed* by Blackett. The equation was *proposed* by Blackett in the following sentence when he wrote about the residual nucleus: "It must however, have a mass-17, and provided no other nuclear electrons are gained or lost in the process, and atomic number 8. It ought therefore, to be an isotope of oxygen." The equation was *confirmed* by Blackett's observations. (Blackett, 1925, 356)

Blackett's results were replicated and his conclusions repeated a year later by William Draper Harkins (1873-1951), a professor of chemistry, and Hugh Allen Shadduck, a graduate student, at the Kent Chemical Laboratory at the University of Chicago. Two years earlier, Harkins and a colleague reported a failure to replicate Wendt and Irion's exploding-wire transmutation experiment. A year earlier, Harkins spoke critically to the news media about the mercury-to-gold transmutation claim of Miethe and Stammreich.

Harkins presented his and Shadduck's paper at the National Academy of Sciences on April 26, 1926. In the paper, published later that year, Harkins marginalized Blackett's work and promoted his own work as the first such achievement. A news story that published a few weeks after the National Academy of Sciences meeting made no mention of Blackett and depicted the discovery as if it belonged to Harkins. (Harkins and Shadduck, 1926)

In 1948, Blackett was awarded the Nobel Prize in physics "for his development of the Wilson cloud chamber method, and his discoveries therewith in the field of nuclear physics and cosmic radiation." With rare exception, Blackett's credit for his discovery of a man-made nuclear transmutation of a stable element has been misattributed to Rutherford.

It is reasonable to credit Rutherford for hypothesizing that change in elemental identity might occur. Some people might even imagine that Rutherford hypothesized the change of nitrogen to oxygen. But scientific credit is not normally given to people who claim they thought of something first but failed to publish.

On the matter of the change of nitrogen to oxygen, Blackett, not Rutherford, experimentally demonstrated and observed the result, obtained the supporting data, and publicly communicated that outcome to the scientific community. Rutherford's accomplishment is the discovery of the change of nitrogen to a proton — and a residual nucleus that he did not correctly identify.

Postscript

In addition to the sources I have cited in this chapter, I also want to acknowledge the helpful research and writing of Bernard Fernandez and Georges Ripka. (Fernandez and Ripka, 2012)

CHAPTER 24

A Bold Advance, A Hasty Retreat

Paneth and Peters Claim Helium by Transmutation, Then Lose Their Courage and Try to Retract (1926-1927)

Our journey through the low-energy transmutation research of the early 20th century concludes with this chapter. Here is where we stand: *Scientific American*, in November 1925, reported its sponsored failure of Adolf Miethe's mercury-to-gold transmutation and, in doing so, reinforced the stigma of alchemy associated with low-energy nuclear transmutation research. On the other hand, Patrick Blackett's high-energy transmutation experiment, which followed Ernest Rutherford's high-energy disintegration experiment, established the credibility of high-energy nuclear transmutations.

The final event in this low-energy transmutation era occurred on Sept. 15, 1926. The players were Friedrich Adolf Paneth (1887-1958), 38, and Kurt Gustav Karl Peters (1897-1978), 29. At the time, the chemists were primarily affiliated with the Chemical Institute of the University of Berlin.

Paneth became known worldwide for his achievements in inorganic and analytical chemistry and was a pioneer in the field of radiochemistry. He was also the founder of a new discipline, cosmochemistry, the study of the chemical composition of matter in the universe, primarily by examination of meteorites.

Paneth was born in Austria and later became a British citizen. He earned a Ph.D. in organic chemistry at the University of Vienna in 1910 and soon shifted his interest to radiochemistry. Early in his career, he was an instructor of inorganic chemistry and radioactivity at the University of Vienna and later became a professor at several universities in Germany. He also worked at the Vienna Institute for Radium Research. When he went to the United Kingdom, he studied with Frederick Soddy at the University of Glasgow as well as with Ernest Rutherford at the University of Manchester.

Friedrich "Fritz" Adolf Paneth (Photo: Max Planck Institute for Chemistry)

After working at a variety of research institutes, he returned to Germany as the director of the Chemistry Department of the Max Planck Institute for Chemistry. According to Silke Merchel, a nuclear chemist who performs research on meteorites at Dresden in Germany and an expert on Paneth's work in cosmochemistry, Paneth not only was interested in pure experimental science but also loved to draw on the secrets of ancient scientists' works. He published more than 50 papers about the history of science and philosophy of chemistry.

Paneth was one of the founders and original editors of the journal *Geochimica et Cosmochimica Acta,* founded in 1950. The F.A. Paneth

Meteorite Trust, administered by the U.K. Royal Astronomical Society, was set up in 1960 in his honor to "encourage and further research concerned with meteorites."

Peters, also born in Austria, worked in the areas of fuel technology, physical chemistry, catalytic reactions, and the separation of rare gases and hydrocarbons. He worked with Paneth in Berlin from 1923 to 1928.

Pumping Hydrogen

Paneth and Peters published their seminal paper in *Berichte der Deutschen Chemischen Gesellschaft*, the Journal of the German Chemical Society. It was reprinted a few weeks later in *Naturwissenschaften*. (Paneth and Peters, 1926, *Berichte der Deutschen*, 1926, *Naturwissenschaften*)

They reported that they had transmuted hydrogen gas into helium. None of their papers appears to have any diagram of their experiment; therefore, I have made a simple conceptual diagram of one of their experimental configurations:

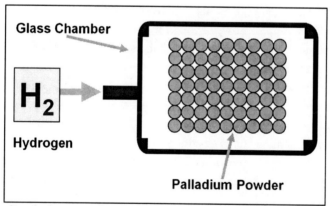

Simple diagram of Paneth-Peters' second experimental method

They placed finely divided palladium in a glass chamber and, without any electrical or heating input, flowed hydrogen gas into the chamber and allowed the palladium to absorb the gas. They typically ran the experiment for hours or a few days. The idea of running an experiment without input energy (aside from the flow of hydrogen) was novel; previous transmutation experiments had always used either electrical discharges or alpha-particle emissions to stimulate the reactions.

In the *Naturwissenschaften* paper, Paneth and Peters expressed the opinion that all of the previous claims of transmuted helium by way of electrical discharges had been "disproved by examinations performed more accurately than the work leading to positive results." In the previous chapters, I have shown that there was no such disproof in those earlier studies. Paneth's attitude toward his peers reveals the recurrence of a common theme: The idealistic pursuit of pure scientific knowledge can be overshadowed by the self-interest of competing scientists when the stakes are high.

Paneth and Peters, of course, were aware of the concern that the helium they observed could have permeated the glass from the atmosphere. They explained their detection limits, and the strength of their signal, and provided reasonable support to counter the argument of contamination. In their paper, they explained that they wanted to try an experiment without any input energy:

> Efforts to realize this element conversion have not been successful up to now, although it has already been tried with various kinds of electrical discharges with large amounts of energy. It can be assumed that the reaction itself delivers very much energy. From the mass loss of the change-over of 4g of hydrogen to helium, a reaction heat of 6.4×10^{11} calories can be calculated. Therefore it is not even certain that any energy at all needs to be added to initiate the reaction
>
> Another opportunity to prove the reaction could be to accelerate catalytically the immeasurably slow conversion of the elements.
>
> The basic idea of our work was therefore to examine whether some hydrogen can convert itself partly to helium without energy supply if hydrogen is brought together with a suitable catalyst. From the beginning, we thought about palladium as the catalyst. (Paneth and Peters, 1926, *Naturwissenschaften*)

They also knew that, if they saw neon in the experiment, along with helium, this would suggest atmospheric contamination. The presence of helium lines within the spectroscope and a total absence of neon lines would have provided very convincing evidence. Unfortunately, because of the spectroscopic sensitivity, they were not able to achieve a total absence of neon, but they came close.

The comparison between the presence of helium and the presence of neon in the experiment provided a qualitative measurement. They also discussed the idea of using a quantitative measurement to determine the presence of transmuted helium, but they did not rely on a quantitative measurement. "We have not evaluated this detection limit in detail," Paneth and Peters wrote, "because in our deciding experiments, we did not need the detection of helium with the absolute limit."

Eureka!

In their initial experiments, they did not use the method depicted above, with pressurized hydrogen, to load the gas into a closed chamber filled with palladium. Instead, they flowed hydrogen through a red-hot capillary of palladium. They noticed that much larger amounts of hydrogen would flow through the capillary when it was hot rather than cold. They "therefore assumed that the generation of helium maybe takes place at the surface. ... If this assumption was correct, the [helium] effect should increase with increased palladium surface."

They switched to a new method to take advantage of this idea. They also began using three types of palladium materials. "We made the next experiments with different preparations of palladium powder, palladium sponge and palladium-asbestos. Instead of flowing hydrogen through palladium, we let the hydrogen simply be absorbed by the palladium and burned the hydrogen after different times with oxygen at the same preparation of palladium." They obtained strong results:

> With this, we got a great simplification of the apparatus and the exclusion of air was much easier than in the experiments with the glowing capillary of palladium.

> Simultaneously with the cleanliness of the helium spectra, the size of the effect increased. We succeeded to produce preparations which, after only 12 hours of heating together with hydrogen, released enough helium to recognize four to five lines of the helium without any line of neon. (Paneth and Peters, 1926, *Naturwissenschaften*)

They began to see a proportional relationship between the helium observations and the duration of the experiments. They also saw two anomalies. First, not all preparations yielded helium. This inconsistency suggests that they were unaware of at least one important parameter that was responsible for the successful experiments. (It also could be argued that a failure condition leading to false positives was acting intermittently.) However, if an inherent flaw in the design resulted in leaks from the atmosphere, they should have seen helium all the time. The second anomaly is that the samples lost their ability to produce helium over time:

> Now it is also known that the capability of palladium to absorb hydrogen has variations which can hardly be explained, and that active palladium has the tendency to become inactive. We have tried the methods recommended to activate palladium again. These are: heating in hydrogen or in oxygen or in a mixture of both or in vacuum. In many cases we were successful; palladium, which no longer generated helium in a significant amount, was after such a treatment, in this way, active again. (Paneth and Peters, 1926, *Naturwissenschaften*)

If the reaction between the hydrogen in the palladium was truly producing helium and consuming the hydrogen, then their process to reactivate the palladium by heating it in hydrogen certainly would have helped restart the reaction because new hydrogen would be adsorbed onto and into the palladium.

They knew that hydrogen loading into the palladium was a requirement. In every case where hydrogen did not sufficiently load into the palladium, they did not see the production of helium. However, some other unknown factor or factors were also required for the production of helium. "On the other hand, preparations which absorbed hydrogen very well, in some cases also generated no or only a small amount of helium, especially in cases when we allowed the hydrogen to absorb under heat."

Fruitless Search for Errors

They asked the question, What if there was no macroscopic leak but the helium permeated the glass from the atmosphere? "As noted above," Paneth and Peters wrote, "a small, incompletely sealed part of the apparatus is insufficient for the explanation, because then helium and neon should be present in a similar proportion as in air. Helium from the air can only be used for the explanation if there is a mechanism for the fractioning of the air which favors the helium [over neon]." In fact, they later found such a mechanism, but they figured out how to compensate for it:

> But under certain circumstances such a mechanism could be present. While glass at room temperature for the duration of our experiments was sufficiently helium tight, at higher temperatures very noticeable amounts of helium went through the glass. Neon passes through the glass to such a smaller extent, that it is possible to generate nearly neon-free helium by flowing air through hot glass.
>
> We therefore made sure that the single part of our apparatus which had to be heated for a short time – the small tube with the palladium preparation to burn the hydrogen and to drive out the helium – was surrounded by a vacuum chamber placed under water. As many calibration experiments showed, the penetration of helium from the air is completely avoided. ...
>
> In our apparatus the amount penetrating through the glass after about a week is just noticeable. In experiments

> carried out over days or weeks, we have taken the precaution to keep the relevant glass parts of the apparatus underwater, with the result with that the amount of helium penetrating within months remains below the limit of our detection method. (Paneth and Peters, 1926, *Naturwissenschaften*)

Only With Hydrogen

They got different results when they used palladium powder rather than palladium sponge or palladium-asbestos. They also tried platinum instead of palladium; they saw some helium, but the magnitude of the effect was much less.

The samples seemed to become inactive over time; that is, they stopped producing helium. They attempted to reactivate the palladium by exposing it to either oxygen or hydrogen gas. Only re-exposure to hydrogen would result in the continued production of helium. This is an essential point. Here are their words from their seminal paper in *Naturwissenschaften*:

> The lack of generation of helium under oxygen exposure and the re-generation under hydrogen exposure seems to not allow another explanation.
>
> Even if one would assume – contrary to what is shown in the previous section – that palladium contained helium, one would need another hypothesis to explain why helium is released only under the influence of hydrogen.
>
> But hydrogen does not have the capability to make palladium permeable for helium; we have proven this in an experiment: a capillary of palladium glowed in a mixture of pure helium (from monazite sand), and hydrogen showed as little permeability for helium as the glowing in helium without hydrogen. (Paneth and Peters, 1926, *Naturwissenschaften*)

They considered searching for heat generation but realized that, based on the amount of helium produced, the amount of heat produced by the reaction would have been too small to detect. They searched for gamma radiation and emissions of ions but found none.

They wrote that, in preparations with high surface area "laying around unused for a longer time at room-temperature – always having some hydrogen bonded – helium should be detectable." I do not understand why they thought helium should be detectable in such cases. Nevertheless, they tested samples of palladium-asbestos that they received from the Kahlbaum company, and they determined that "this palladium-asbestos is still unusually active for helium production."

They knew that palladium was not permeable for helium, but perhaps the behavior of palladium-asbestos was different. Regardless, Paneth and Peters again found that experiments performed in oxygen produced no helium and experiments performed in hydrogen produced significant amounts of helium:

> After a 12-hour exposure with oxygen, there was a minimally detectable amount of helium, which can easily be explained by the residual hydrogen. After a five-hour exposure with hydrogen immediately afterwards, the analysis resulted in 10 [to] 100 times the amount of helium.
>
> We repeated the experiment three times sequentially – always with the same result. When we started with the examination of this palladium-asbestos, the helium generation was 10^{-8} to 10^{-7} cm^3/day; After 20 cycles of heating, even this palladium-asbestos became inactive. As a result of several subsequent oxygen and hydrogen treatments at various temperatures, it again became active for helium generation, but not to the same extent as in the beginning. Per day it only delivered 10^{-9} to 10^{-8} cm^3 of helium, an amount also reached by some preparations of palladium-asbestos produced by us. (Paneth and Peters, 1926, *Naturwissenschaften*)

When *Nature* summarized Paneth and Peters' *Deutschen Chemischen Gesellschaft* paper a few weeks later, *Nature* repeated Paneth and Peters' idea that somehow palladium preparations that had sat unused at room temperature should contain a little helium. Again, I do not understand why those preparations should have contained any helium. (*Nature*, 1926, 526, Reported Conversion)

The First Real Alchemist?

Forty-eight hours after Paneth and Peters' *Deutschen Chemischen Gesellschaft* paper published, the news hit the papers. On Sept. 17, 1926, Reuters wrote and distributed the story. In the U.K., the *Morning Post* picked up the Associated Press story, and on Sept. 25, 1926, *Nature* picked up the *Morning Post* story but mentioned the news only in very general terms. (*Nature*, 1926, News and Views)

Nature made a key point about the Paneth and Peters claim:

> This announcement, if correct, is of great importance and will evoke even more interest than the claim by Miethe and Stammreich to have transmuted mercury into gold. The two claims differ, however, in the important respect that, whereas the experiments of Miethe and Stammreich, and of Smits, indicated the disintegration of heavy atoms into lighter ones, those now announced [by Paneth and Peters] involve the synthesis of an element from a lighter one, thus more nearly approaching the alchemists' dream.

Paneth was eager to establish himself as the first successful alchemist. He sailed to the United States; his destination was Cornell University, in Ithaca, New York, where he began a series of lectures as a non-resident lecturer. At 8 p.m. at Cornell's Baker Laboratory lecture hall on Oct. 4, 1926, Paneth gave his first lecture; "Ancient and Modern Alchemy."

> **DEVELOPMENT OF ALCHEMY TRACED BY FRITZ PANETH**
>
> Ancient and Modern Conceptions of Science Contrasted by Speaker
>
> **ANCIENT EFFORTS FUTILE**
>
> Men Were Chiefly Interested in Transmutation of Baser Metals to Gold
>
> **POINT OF VIEW CHANGED**
>
> Scientists Endeavor Today to Analyze the Atomic Correlation of Matter

Headline from Oct. 5, 1926, news story in Cornell Daily Sun

The text of Paneth's first lecture at Cornell was published in *Science* three days later. (Paneth, 1926, Ancient) He spent a great deal of time discussing the work of the ancient alchemists, then moved into a discussion about Ernest Rutherford's 1919 report of the disintegration of nitrogen and emission of a proton. Paneth, after explaining the details of the Rutherford experiment and its reliance on optical data "so feeble that only the trained eye can perceive it under the most favorable conditions," made an interesting judgment about Rutherford. "We must therefore marvel," Paneth wrote, "at the boldness of Rutherford, who ventured to draw revolutionary conclusions from this apparently negligible phenomenon."

Rapid Retreat

On Dec. 9, 1926, the Paneth and Peters story hit the Associated Press news service in the U.S. Exactly two months later, on Feb. 9, 1927, the two chemists began their retreat. (Paneth, Peters, and Günther, 1927) They submitted the results of further research to *Berichte der Deutschen Chemischen Gesellschaft*. (Thanks to Maxime Petiau for help with the translation.)

Paneth had been doing more research at Cornell with his colleague Paul L. Günther, and Peters had been working in Berlin. They wrote that they had "underestimated the influence of two sources of errors." First, they wrote that they believed that traces of helium pre-existed in the palladium-asbestos:

> Asbestos contains, as all minerals, traces of helium, and as such, we took the precaution to strongly heat up the asbestos used in the preparation of the palladium-asbestos so that, by the weaker heating during the experimentation, it would release no more helium. To clarify the behavior of the palladium-asbestos of Kahlbaum, we tried an experiment with non-preheated asbestos and established surprisingly that it released its helium in hydrogen at lower temperature than in oxygen. (Paneth, Peters, and Günther, 1927)

Here is the text of their footnote to that paragraph on which they based their conclusion that asbestos contained traces of helium: "In the literature, we did not find any report on the helium content of asbestos; according our experiments, it can be up to 10^{-4} cm^3 per gram."

I do not know whether subsequent studies on asbestos have confirmed that it naturally contains helium. In any case, if the palladium-asbestos "released" more helium — one-to-two orders of magnitude more — in the presence of hydrogen than it did in the presence of oxygen, that *confirms* rather than *disconfirms* their claim. They explained their thinking further:

Therefore, as the Kahlbaum palladium-asbestos released helium under a hydrogen load, but almost none under an oxygen load, we don't doubt that that helium was released from the asbestos. The connection between the hydrogen activity of the palladium and the release of the helium, and the assumption of the origin of the helium in palladium is indirect: *Only when the palladium was activated and had absorbed hydrogen, and the asbestos was immersed in a hydrogen atmosphere, and only then the helium was released.* [Emphasis added] (Paneth, Peters, and Günther, 1927)

Next, Paneth and Peters speculated about how they could have made a mistake in their experiments that used material other than the Kahlbaum palladium-asbestos:

If this explanation [pre-existing helium in asbestos] for the presence of helium is correct, then the special experimental result on which we placed emphasis is to be canceled, and only the experiments with our own preparations remain. These are preparations with no asbestos (palladium sponge and palladium powder) or very strongly heated helium-free asbestos. In those cases this source of error cannot have taken place; we think a different source of error was underestimated in its effect. (Paneth, Peters, and Günther, 1927)

They explained that, although they had previously accounted for the possibility of occluded helium in the glassware, they underestimated the quantity of glass used:

We believe that this source of error — release of the helium absorbed in glass through the vacuum chamber and intrusion of the helium through the glass wall of the preparation tube in the analysis equipment — is responsible in many of our experiments for the presence of [the] helium. ... The way we think the helium intruded into the equipment will be made clear when the drawings from the equipment will be published. (Paneth, Peters, and Günther, 1927)

As far as I can tell, they never published drawings. Despite their attempts to explain away their claim of helium production in each and every one of their configurations, they were unable to do so. They assumed that a rational explanation for their remaining apparent positive results would eventually present itself:

> For the rest of the positive tests, even today, we cannot give an explanation. But since the majority of our experiments have explained themselves in a "natural" way, we think it probable that it will also happen for our outstanding (unexplained up to now) experiments. (Paneth, Peters, and Günther, 1927)

Perplexing Explanation

On March 2, 1927, Paneth (without Peters) submitted an English synopsis of his retraction to *Nature*. It includes a little more information and is generally consistent with the translation I obtained from the German retraction. In this paper, Paneth summarized his perspective more concisely, though not entirely coherently:

> We feel that we are in a position to give an explanation of the occurrence of the observed *very small quantities of helium* in our experiments, without having recourse to the assumption of a synthesis of helium.
>
> [In earlier papers, we] discussed the possibility of regarding the helium dissolved in the glass as an explanation of the observed effects, but blank experiments led us to the conclusion that the quantity of helium capable of being liberated in this way was beyond the limits of sensitivity of our method of detection.
>
> [Since then], we have carried out experiments both in

the Baker Laboratory of Cornell University and in the Chemical Laboratory of the University of Berlin, and these have shown that the liberation of helium from glass (and from asbestos) is dependent on the presence of hydrogen. *Thus glass tubes which gave off no detectable quantities of helium when they were heated in a vacuum or in oxygen were found to yield helium in quantities on the order of 10^9 cc when they were heated in an atmosphere of hydrogen.* [Emphasis added]

Now, in the earlier experiments, the glass tubes containing palladium yielded helium, whereas the empty glass tubes used in control experiments did not; and since the former tubes would fill [were filled?] with hydrogen on the application of heat, we see that the source of the helium lay not in the palladium but in the glass, in spite of appearances to the contrary. ...

Since asbestos behaves similarly to glass [helium absorption and release], we now see why one particular palladium preparation, bought as palladium-asbestos, yielded larger quantities (10^{-7} c.c.) of helium after being charged with hydrogen. Here, obviously, in contrast to the preparations we made ourselves, the asbestos had not been ignited [activated?] until it was free from helium, and a fraction of the residual helium was always liberated by heating when the palladium was charged with hydrogen, whereas in oxygen, no development of helium could be observed. (Paneth, 1927, *Nature*)

Thus, they had conducted tests with asbestos as well as palladium-asbestos and found that, using both materials, more helium was "released" in the presence of hydrogen than in the presence of oxygen or a vacuum. They offered no explanation for why this should happen. This represented a major omission. In their own words from their September paper, they wrote that "one would need another hypothesis to explain why helium is released only under the influence of hydrogen." They did not provide such an explanation.

Nevertheless, I tried but failed to find an explanation for helium being released only under the influence of hydrogen. I found that, in the book *Too Hot to Handle: The Race for Cold Fusion*, particle physicist and science popularizer Frank Close referred to this anomaly as *support* for Paneth's retraction.

Frank and I have gotten to know each other over the past few years, and we have discussed and debated several parts of "cold fusion" history. In an e-mail, I asked him, "Why would glass permeate helium in an atmosphere of hydrogen but not in oxygen?" He replied that it "may be linked to the fact that oxygen atoms are heavier than helium, whereas hydrogen is lighter, and the basic mechanical argument is that evaporation is easier in hydrogen than oxygen." But Paneth and Peters observed no helium when they performed the experiments in the presence of a vacuum, so that hypothesis doesn't work.

Helium does not permeate intact, defect-free metal. Unless it is created within metal, it cannot be released from metal without extreme mechanical or thermal stimuli. Thus, the qualitative determination of the presence of helium — only when the palladium was loaded with hydrogen — means either that Paneth and Peters produced helium in a hydrogen-palladium reaction or, by some miracle, hydrogen has the unique ability to draw helium through glass.

Paneth and Peters also attempted to *quantitatively* explain away their helium with a revised interpretation of their detection limit. I do not know whether this new revision would withstand rigorous scrutiny. Nevertheless, in their attempt to explain away their *qualitative* determination, they in fact reinforced the fact that heated glass tubes containing palladium, only when they were filled with hydrogen, produced helium.

Perplexing Reversal

I don't understand the sudden shift in Paneth's mind. Paneth and Peters worked on their experiments for three years. They were in no rush to publish. I know of no other close competitors who attempted a similar method at that time. They had lots of time to consider how to carefully set up their experiments, analyze their data, and write their paper. And indeed, their paper was thorough, thoughtful and detailed, though it did lack diagrams.

Immediately after publication of the paper, Paneth went to Cornell and proclaimed that he had succeeded where modern transmutation claimants and ancient alchemists had failed. He published the same bold assertion in *Science* a few days later. Paneth was neither timid nor reserved. Conversely, in his Cornell lecture, Paneth spoke critically about Rutherford, for his boldness in drawing "revolutionary conclusions from [an] apparently negligible phenomenon."

Four months later, Paneth and Peters published their retraction paper in the *Journal of the German Chemical Society,* followed by a similar retraction in *Nature.* Both of these retractions were ambiguous and — unintentionally, I presume — reinforced the point that helium was produced only in the presence of palladium and hydrogen. Paneth struggled to explain away each of the various tests but failed. End of story.

Within a few years, Paneth took an interest in the new field of high-energy physics and used neutrons (the existence of which was confirmed in 1932 by James Chadwick) to pursue nuclear transmutations. In 1935, Paneth claimed that he and his co-author H. Loleit were the first people to achieve an alchemical transmutation of sufficient magnitude to be measured chemically. (Paneth and Loleit, 1935)

So why did Paneth make such a rapid retreat? I have searched the news archives for evidence of conflict and criticism. With the exception of one newspaper report of an attempted and failed replication at Princeton University, I have found no evidence of controversy between the time Paneth published his claim and when he retracted it. Paneth's about-face is inexplicable to me. I am convinced that something else happened to impel his retreat, but I do not what that was.

Additionally, there is a parallel to Paneth's behavior. Fritz Haber, who in 1925 had affirmed the transmutation claims of Adolf Miethe and Hantaro Nagaoka with great confidence, also retreated from his position with dubious explanations. Haber retreated in May 1926, just months before Paneth and Peters made their bold claim.

After Paneth and Peters, low-energy nuclear transmutation research was dormant for 60 years.

CHAPTER 25

Asleep for 60 Years

Summary of the New or Forgotten Contributions to the Scientific Literature Shown in This Book

By 1930, low-energy elemental transmutation research was dormant. In its place, research in high-energy physics took the spotlight. Within a decade, thermonuclear fusion and nuclear fission were discovered.

In 1932, scientists began using devices to accelerate ions and induce nuclear disintegrations. These devices provided much higher energies and greater control over beam intensities than offered by alpha particles randomly emitted from passively decaying radioactive elements.

The dormancy of low-energy nuclear transmutation research is no mystery. High-energy nuclear transmutations were easier to understand and describe with theory, and the experiments could be repeated and controlled. Low-energy transmutations conflicted with accepted theory and were difficult to repeat and control.

Contributions to Scientific Literature

The following list summarizes the new or forgotten contributions to the scientific literature from the early 20th century that are revealed in this book.

1907 — Ramsay's Unconfirmed Transmutation Claim

The lost era of low-energy nuclear transmutations began in 1907 with the work of William Ramsay. He had been the first scientist to use the energy of alpha particles to attempt man-made nuclear transmutations. He placed various dissolved compounds in water and subjected the solutions to alpha particle emission. Nobody seems to have replicated and confirmed his claims. Neither is there any unambiguous disproof of his claims.

1912-1914 — High-Voltage Discharges

Beginning in 1912, several independent researchers reported results from experiments they had performed using glass cathode-ray tubes fitted with electrodes, in a vacuum or filled with hydrogen. The researchers ran continuous, high-voltage electric discharges through the tubes for hours or days, then analyzed the residual gases with spark optical emission spectroscopy. They detected the production of the gases helium-4, neon, argon, and an element with an atomic mass of three, either tritium or, less-likely, helium-3. The researchers performed a wide variety of tests to rule out simple explanations. Nobody could make any sense of the results at the time.

Researchers reporting some of these results included William Ramsay, John Norman Collie, Hubert Sutton Patterson, Joseph John Thomson, James Irvine Orme Masson and George Winchester. Some of them denied their own results and claimed that the rare gases leaked into their apparatus or were inexplicably embedded into components of the apparatus before the experiments began. Neither speculation was confirmed or is well-supported by fact.

Although critics offered many guesses in their attempts to dismiss the claimed results, none identified a specific error of protocol, a mistake in the data analysis, or an unstated assumption by these researchers. Many historians have casually dismissed these experiments merely because other scientists failed in their replication attempts.

This collection of experiments demonstrated a broad, independent basis for the first confirmed man-made nuclear transmutations. These 1912-1914 transmutations were produced by low-energy processes rather than by high-energy alpha bombardment.

1919 — Rutherford's Man-Made Disintegration

In 1919, Ernest Rutherford used the energy of alpha particles to initiate a man-made disintegration of a stable element. He showed that when an alpha particle impacts a nitrogen atom, a proton is ejected. The distinction between hydrogen (the element) and protons (the subatomic particles) was not clear at the time. Therefore, some people mistakenly attributed Rutherford's disintegration to a transmutation of the element nitrogen to the element hydrogen, which is incorrect.

1922 — Exploding Wires Produce Helium

Gerald L. Wendt and Clarence E. Irion, using the exploding electrical conductor phenomenon, synthesized helium, in a split-second burst of high-voltage electrical current.

Between the time the chemists presented their paper at the American Chemical Society meeting and when their paper published, Rutherford publicly criticized their claims. He did so in a vacuum of information, and Wendt showed that it was technically irrelevant. Nevertheless, because of Rutherford's stature, his commentary strongly influenced public and scientific opinion. Neither Rutherford nor anyone else unambiguously identified any error of protocol, measurement, analysis, or unstated assumption in Wendt and Irion's 21 successful experiments.

There were no serious replication attempts in the 20th century, but this is understandable because Rutherford aggressively but unscientifically critiqued the Wendt-Irion claims without waiting for their published paper.

At about the same time, a news article in a major Paris newspaper proclaimed Rutherford as the first modern scientist to accomplish a transmutation of the elements. He was not; the news story was an erroneous depiction of Rutherford's 1919 disintegration of nitrogen.

1924-1925 — Synthetic Gold Claims, Germany

In 1924, German scientist Adolf Miethe accidently found trace amounts of gold in the residue of mercury vapor lamps that he had been using for photography. He claimed that he had transmuted mercury to gold and possibly platinum in the high-powered electric lamps. Gold and platinum are one and two steps down, respectively, on the periodic table of elements from mercury.

Arthur Smits and Albert Karssen, in Amsterdam, followed the general method described by Miethe, but instead of starting with mercury, began with lead. They observed the production of mercury and the rare element thallium. Thallium and mercury are one and two steps down the periodic table, respectively, from the element lead.

Scientific American magazine sponsored research by scientists at New York University to replicate the Miethe experiment. After the New York experiments failed to produce positive results, the magazine deemed Miethe's experiment nothing but a fantasy simply because that group failed in its replication attempt.

A prominent German chemist, Fritz Haber, contributed to the idea that the positive results had been in error. He speculated that the cathodes used in Miethe's experiments contained gold before the experiment began. No critics unambiguously identified a specific error of protocol, a mistake in the data analysis, or an unstated assumption in his experiments. Milan Wayne Garrett, at the Clarendon laboratory of Exeter College, Oxford University, tested Haber's idea. It didn't work. Despite reports of four replications — the University of Amsterdam, the Institute of Physical and Chemical Research in Japan, Siemens, and the Kent laboratory at the University of Chicago —interest in these gold-producing low-energy transmutation experiments died.

1924-1925 — Synthetic Gold Claims, Japan

Between 1924 and 1925, Japanese scientist Hantaro Nagaoka discovered specks of gold while he was examining spectroscopy lines of mercury and bismuth. He observed the gold after applying a high electric field to mercury, using tungsten electrodes, in the presence of a hydrogen-bearing material: transformer oil. Nagaoka also reported observing a second substance, which he described as having the appearance of platinum.

Despite extensive English-language searches, I found no evidence that Nagaoka's results were challenged. Although Nagaoka traveled around the world displaying his synthetic gold, there is no evidence that anyone tried to repeat his experiments.

1925 — Blackett's Nitrogen-to-Oxygen Transmutation

In 1925, a nuclear transmutation occurred along a different, more-understandable line of research. Patrick Maynard Stewart Blackett, a research fellow working under Ernest Rutherford, used alpha particle bombardment to show the transmutation of nitrogen to oxygen. All but a few historians, and all Internet references at the present time, incorrectly credit this discovery to Rutherford.

1926 — Paneth and Peters' Hydrogen-to-Helium Transmutation

In 1926, German chemists Friedrich Adolf (Fritz) Paneth and Kurt Gustav Karl Peters pumped hydrogen gas into a chamber with finely divided palladium powder and reported the transmutation of hydrogen into helium.

In 1927, Paneth submitted a partial retraction of his and Peters' claims. Paneth asserted that the helium had come from leaks from atmospheric air. His retraction, which included the following statement, was internally inconsistent: "Thus glass tubes which gave off no detectable quantities of helium when they were heated in a vacuum or in oxygen were found to yield helium in quantities on the order of 10^{-9} cc when they were heated in an atmosphere of hydrogen."

Beyond Paneth's aggressive attempts to deny his and Peters' own work, no critics identified a specific error of protocol, a mistake in the data analysis, or an unstated assumption in their experiments.

Considering the theoretically inexplicable results, the difficulty in performing the experiments, and certain backlash from more-conservative scientists, other scientists had little incentive to attempt replications.

If *Lost History* teaches anything, one lesson is paramount: A failure to replicate by one scientist has no direct bearing on the successful claim of another scientist.

Developments in Nuclear Physics

Soon after James Chadwick experimentally confirmed the existence of the neutron in 1932, scientists discovered the concept of the nuclear chain reaction. After that came the atomic fission bomb, then the hydrogen bomb, and eventually, the peaceful use of the atom for nuclear power, nuclear medicine and other purposes.

The idea of low-energy-based nuclear reactions remained dormant for 60 years. The disappearance of this science for 60 years may seem unbelievable, but it is, in fact, what happened. I want to again acknowledge the only other person I know of who has recognized and researched this forgotten era of science: Robert Nelson.

Fusion Fiasco:
Explorations in Nuclear Research, Vol. 2

In 1984, two chemists began secretly working in the basement of the Chemistry Building at the University of Utah, attempting to make nuclear reactions with chemistry. They were well aware of the risks to their careers, among other factors, with such a pursuit.

In 1989, they were forced by circumstances largely beyond their control to go public. They claimed that they had found a new route to nuclear fusion. Just like Abel Niépce de Saint-Victor, who correctly observed radiation in 1858 but incorrectly interpreted the phenomena, the Utah chemists too observed valid anomalous phenomena and made an incorrect interpretation of it.

The Utah chemists' announcement ignited one of the biggest science controversies of the 20th century.

Hacking the Atom:
Explorations in Nuclear Research, Vol. 1.

From 2000 to 2012, as a journalist, I investigated the research in the low-energy nuclear reactions (LENRs) field, initially and erroneously called "cold fusion." These developments, as well as the promising non-fusion theory published in 2006 by Allan Widom and Lewis G. Larsen, compose *Hacking the Atom: Explorations in Nuclear Research, Vol. 1.*

The Widom-Larsen theory, with no close runner-up, best explains the experiments conducted between 1912 and 1927 as well as most of the phenomena reported in LENRs in the last 26 years.

Appendix A – Emissions From Spontaneous Radioactivity

Type	Nature of Radiation	Penetrating Power[1]	Ionizing Power [2]
Alpha α $^{4}_{2}He$	A helium nucleus of 2 protons and 2 neutrons. Mass = 4 Charge = +2	**Low** Particles are stopped by a few cm of air or a thin sheet of paper.	**Very High** The biggest mass and charge of the three. Packs the biggest punch.
Beta β $^{0}_{-1}e$	High kinetic energy electrons. Mass = 1/1850 Charge = -1	**Moderate** Most particles are stopped by a few mm of metals like aluminum.	**Moderate** Less than the alpha particle.
Gamma γ $^{0}_{0}\gamma$	Very high frequency electromagnetic radiation. Mass = 0 Charge = 0	**Very High** Most, but not all, gamma rays are stopped by a thick layer of steel or concrete, or a few cm of dense lead.	**Lowest** Carries no electric charge and has no mass, so it has very little punch when it collides with an atom.

1. When penetrating denser material, more radiation is absorbed and stopped than when penetrating less-dense material. However, as mass or charge decreases, the penetrating power increases.
2. Ionizing power is the ability to remove electrons from atoms and form positive ions. Ionizing radiation is harmful to living cells. Courtesy Georgia State University, adapted by *New Energy Times*.

Appendix B - Helium Permeation in Metals Analysis

Gas Behavior in Hydride-Forming Metals at or Near Standard Temperature and Pressure	Hydrogen	Helium
Readily permeates hydride-forming metals	Yes	No
Diffuses through defects, cracks or grain boundaries in metals	Yes	Yes
Soluble (dissolves) in hydride-forming metals	Yes	No

Bowman Jr., Robert C. (Feb. 7, 2007) "NMR Studies of 3He Retention and Release in Metal Tritides — A Review," Hydrogen & Helium Isotopes in Materials Conference, Albuquerque, N.M. *[Helium does not outgas from metals easily or quickly.]*

Chien, Chun-Ching, Hodko, Dalibor, Minevski, Zoran and Bockris, John O'M. (April 1992) "On an Electrode Producing Massive Quantities of Tritium and Helium," *Journal of Electroanalytical Chemistry*, **338**, 189-212 *[Helium on near-surface areas on cathode can be retained if quickly immersed in liquid nitrogen.]*

Gozzi, D., Cellucci, F., Cignini, P.L., Gigli, G., Tomellini, M., Cisbani, E., Frullani, S., Urciuoli, G.M. (1998) "X-Ray, Heat Excess and 4He in the D:Pd System," *Journal of Electroanalytical Chemistry*, **452**, 253, and Erratum, **452**, 251-71 *[Helium does not show up in the bulk if the cathode is vaporized.]*

McKubre, Michael, et al., (June 1998) "Development of Energy Production Systems from Heat Produced in Deuterated Metals, Volume 1," Electric Power Research Institute, TR-107843 *[Researchers hypothesized, but did not test, that helium was retained (occluded) in metal during experiment.]*

Ramsay, W., and Travers, M.W. (January 1897) "An Attempt to Cause Helium or Argon to Pass Through Red-Hot Palladium, Platinum, or Iron." *Proceedings of the Royal Society of London (1854-1905)*, 61(-1), 266-7 *[Helium won't dissolve in metal even at high temperature.]*

Schultheis, D. (2007) "Permeation Barrier for Lightweight Liquid Hydrogen Tanks," Ph.D. dissertation, University of Augsburg *[Defect-free metal will not allow helium to pass through.]*

Xia, Ji-xing, Hu, Wang-yu, Yang, Jian-yu, and Ao, Bing-yun (2006) "Diffusion Behaviors of Helium Atoms at Two Pd Grain Boundaries," *Transactions of Nonferrous Metals Society of China*, 16, S804-7 *[Helium has low solubility in metals, grain boundaries support permeation.]*

Appendix C - Cathode-Ray Tube Research/Rare Gases

Listed by Publication Date

Ramsay, William (July 18, 1912) "Experiments With Cathode Rays," *Nature*, **89**(2229), p. 502

Thomson, Joseph John (Feb. 13, 1913) "On the Appearance of Helium and Neon in Vacuum Tubes," *Nature*, **90**(2259), p. 645-7; also in *Science*, **37**(949), p. 360-4, (March 7, 1913); also in *Scientific American Supplement*, **75**(1940), p. 150, (March 8, 1913)

Nature (Feb. 13, 1913) "Origins of Helium and Neon," **90**, p. 653-4

Ramsay, William (Feb. 15, 1913) "The Presence of Helium in the Gas From the Interior of an X-Ray Bulb," *Proceedings of the Chemical Society*, **29**(410), p. 21-2

Thomson, Joseph John (March 8, 1913) "The Birth of the Atom," *Scientific American Supplement*, **75**(1940), p. 154-5

Thomson, Joseph John (May 29, 1913) "Some Further Applications of the Method of Positive Rays," *Nature*, **91**(2774), p. 333-7

Ramsay, William (published sometime before June 28, 1913) "The Presence of Helium in the Gas From the Interior of an X-Ray Bulb," *Journal of the Chemical Society Transactions*, **103**, p. 264-6

Collie, John N., and Patterson, Hubert S. (published sometime before June 28, 1913) "The Presence of Neon in Hydrogen After the Passage of the Electric Discharge Through the Latter at Low Pressures," *Journal of the Chemical Society Transactions*, **103**, p. 419-26

Collie, John N., and Patterson, Hubert S. (June 28, 1913) "The Presence of Neon in Hydrogen After the Passage of the Electric Discharge Through the Latter at Low Pressures. Part II," *Proceedings of the Chemical Society, London*, **29**(418) p. 217-21

Soddy, Frederick (1913) "Radioactivity," *Annual Reports on the Progress of Chemistry*, Chemical Society, **10**, p. 262-88

Masson, James Irvine Orme (June 28, 1913) "The Occurrence of Neon in Vacuum-Tubes Containing Hydrogen," *Proceedings of the Chemical Society, London*, **29**(418), p. 233

Soddy, Frederick (Dec. 4, 1913) "Intra-Atomic Charge," *Nature*, **92**, p, 399-400

Strutt, Robert John (1914) "Attempts to Observe the Production of Neon or Helium by Electric Discharge," *Proceedings of the Royal Society of London*, **89**, p. 499-506

Winchester, George (April 1, 1914) "On the Continued Appearance of Gases in Vacuum Tubes," *Physical Review*, **3**(4), p. 287-94

Merton, Thomas R. (Aug. 1, 1914) "Attempts to Produce the Rare Gases by Electric Discharge," *Proceedings of the Royal Society of London*, **90-A** (621), p. 549-53

Collie, John N. (Aug. 1, 1914) "Note on the Paper by T. R. Merton on 'Attempts to Produce the Rare Gases by Electric Discharge,'" *Proceedings of the Royal Society of London,* **90-A** (621), p. 554-6

Collie, John N., Patterson, Hubert S., and Masson, Masson, James Irvine Orme (Nov. 2, 1914) "The Production of Neon and Helium by the Electrical Discharge," *Proceedings of the Royal Society of London,* **91-A** (623), p. 30-45

Egerton, A.C.G. (March 1, 1915) "The Analysis of Gases After Passage of Electric Discharge." *Proceedings of the Royal Society of London,* **91-A** (627), p. 180-9

Piutti and Cardoso (1920) *J. Chim. Phys.*, **18**, p. 81

Piutti, Zeitsch, **28**, p. 42 (1922)

Bibliography of Scientific Literature Relating to Helium, 2nd Ed. (Dec. 21, 1922) Bureau of Standards, No. 81, p. 12

Allison, Samuel King and Harkins, William Draper (April 1924) "The Absence of Helium From the Gases Left After the Passage of Electrical Discharges: I, Between Fine Wires in a Vacuum; II, Through Hydrogen; and III, Through Mercury Vapor," J. Amer. Chem. Soc., 46(4) p. 814-24

Riding, R. W. and Baly, E.C.C. (Sept. 1, 1925) "The Occurrence of Helium and Neon in Vacuum Tubes," *Proceedings of the Royal Society of London,* **109-A** (749) p. 186-93

Baly, E. C. C. and Riding, R.W. (Oct. 30 1926) "The Occurrence of Helium and Neon in Vacuum Tubes," *Nature,* **118** (2974,) p. 625-6

Lawson, R.W. (1926) "The Occurrence of Helium and Neon in Vacuum Tubes," *Nature,* **118**, p. 838-9

Appendix D - Rutherford's Nitrogen Disintegration

Listed by Publication Date

Rutherford, Ernest (June 1919) "Collisions of Alpha Particles With Light Atoms. I. Hydrogen," *Philosophical Magazine*, Series 6, **37**(222), p. 537-61

Rutherford, Ernest (June 1919) "Collisions of Alpha Particles With Light Atoms. II. Velocity of the Hydrogen Atom," *Philosophical Magazine*, Series 6, **37**(222), p. 562-71

Rutherford, Ernest (June 1919) "Collisions of Alpha Particles With Light Atoms. III. Nitrogen and Oxygen Atoms," *Philosophical Magazine*, Series 6, **37**(222), p. 571-80

Rutherford, Ernest (June 1919) "Collisions of Alpha Particles With Light Atoms. IV. An Anomalous Effect in Nitrogen," *Philosophical Magazine*, Series 6, **37**(222), p. 581-7

Rutherford, Ernest (1920) "Nuclear Constitution of Atoms" [Bakerian Lecture], *Proceedings of the Royal Society of London* **97**-A, 374-400; reprinted in *The Collected Papers of Lord Rutherford of Nelson O.M., F.R.S. Published under the Scientific Direction of Sir James Chadwick, F.R.S.*, Chadwick, James, ed., **3**, George Allen and Unwin, p. 14-38, (1965)

Rutherford, Ernest (May 13, 1922) "Artificial Disintegration of the Elements," *Nature*, **109**(2741), p. 614-16

Rutherford, Ernest and Chadwick, James (August 1924) "Further Experiments on the Artificial Disintegration of Elements," *Philosophical Magazine*, **36**, p. 417-22

Rutherford, Ernest and Chadwick, James (March 29, 1924) "The Bombardment of Elements by Alpha Particles," *Nature*, **113**, p. 457

Rutherford, Ernest (1925) "Studies of Atomic Nuclei," in *Proceedings of the Royal Institution Library of Science*," **9**, p. 73

Rutherford, Ernest (April 4, 1925) "Disintegration of Atomic Nuclei," *Nature*, **115**, p. 493-4

Appendix E - Timeline of Key Events in Early Nuclear History

This timeline covers key events in nuclear history, with a specific focus on those events that relate to early nuclear transmutation. The preceding chapters have covered a few of the highlights from this timeline.

The timeline covers both well-known discoveries in nuclear physics and two brief, relatively unknown periods when researchers observed anomalous transmutations. The first of these periods began in 1912 and ran until World War I began, in 1914. Researchers used primarily a high-current electrical discharge method. In the second period, from 1922 to 1926, researchers used a variety of experimental methods.

In these periods, at least three dozen papers, published in top-tier journals, chronicle researchers' attempts at man-made transmutations of stable elements without the use of radioactive seed elements or particle accelerators.

Much credit for preserving the history of these lost periods goes to an unaffiliated science enthusiast named Robert Nelson (see Chapter 9) who spent countless hours in the late 1960s at the University of California, Berkeley library, poring through this history. In all likelihood, there are even more important scientific papers from these forgotten periods.

I am grateful for help on archival searches for journal articles and news clips to reference librarian Randy Souther, at the Gleeson Library, Geschke Center, University of San Francisco, Lorna Whyte, senior library assistant at the San Rafael public library, and to Lorna Lippes, a independent researcher.

Two other sources were helpful. The first was the online bibliography "Selected Classic Papers," maintained by Carmen Giunta, professor of chemistry at Le Moyne College. The second was the "Chronology of Milestone Events in Particle Physics," written by an international consortium of high-energy physicists. It is available online at *http://web.ihep.su/owa/dbserv/intr.page1*. A few words of wisdom from their introduction apply equally to my timeline.

"Paths at the frontier of science are rarely straight," the authors wrote. "Along the way, brilliant insights, new experimental tools, and hard-won new data are accompanied by confusion, wrong turns, conservative dogma, or wrong experiments."

With this in mind, the timeline below includes not only successes but also failures and ambiguous results, which were significant in their own right.

Timeline of Key Events In Early Nuclear History

~600 B.C.E.-1800s
Various schools of alchemy, with roots in Egypt, India and China, propose that elements can be transmuted from one to another. (Trenn, 1981)

1808
John Dalton proposes that atoms are composed of a unique material for each element, that each atom is the smallest unit, that atoms are indivisible, and that transmutation (alchemy) of elements is impossible. Dalton's view is broadly accepted among scientists until 1902, when Ernest Rutherford and Frederick Soddy prove otherwise. (Dalton, 1808)

1816
William Prout proposes that all elements are made of different configurations and quantities of hydrogen atoms. He calls it the "protyle" concept. The concept is replaced by the proton in 1920. (Redgrove, 1922)

1858
Abel Niépce de Saint-Victor is the first to observe and record spontaneous radioactive emissions on photographic plates. (Habashi, 2001)

1895
William Ramsay co-discovers the first of the noble gases, argon. (Strutt and Ramsay, 1895)

1895 (November 8)
Wilhelm Roentgen discovers a new kind of ray, which he calls the X-ray. (Roentgen, 1896)

1895
Ramsay discovers helium on Earth by treating the mineral cleveite with mineral acids. Helium had been discovered on the surface of the sun in 1868. (Ramsay, 1895)

1896 (March 3)
Henri Becquerel is the second to observe and record spontaneous radioactive emissions on photographic plates. He uses naturally fluorescent minerals and uranium salts in his experiments. See Rutherford 1898. (Becquerel, 1896)

1897 (August)
Joseph John Thomson, experimenting with cathode rays, discovers the electron. This is one of the first significant discoveries that disproved Dalton's idea of the indivisible atom. (Thomson, 1897)

1898 (July)
Marie Sklodowska Curie and Pierre Curie, working with uranium, observe the production of a new substance, identify polonium and coin the term "radioactivity." (Curie, 1898)

1898 (March)
Ramsay discovers more noble gases: krypton, neon and xenon. (Travers, 1928)

1898 (October)
Rutherford, a student of Thomson, examines the radiation emitted from uranium that Becquerel had reported in 1896. Rutherford distinguishes two types of rays, which he names "alpha" and "beta." They are later identified as helium atoms and electrons, respectively. (Rutherford, 1899)

1898 (December)
Marie and Pierre Curie and Gustave Bémont, while doing experiments with uranium, observe a new substance; they discover radium. (Curie, Curie, and Bémont, 1898)

1899 (December)
Paul Villard observes a new kind of radiation from uranium that is more penetrating than alpha or beta particles. (Gerward, 1899; Villard, 1900)

1900 (December)
Rutherford introduces the concept of radioactive half-life and refers to the time when the radioactivity of an element falls to "one-half its value." (Rutherford, 1900)

1901
Rutherford names the radioactive emission observed by Villard a "gamma ray." (Rutherford, 1903)

1901 (December)
Soddy and Rutherford examine the emanation spontaneously emitting from thorium. They show that, because the emanation was physical matter, as opposed to a wave or energy, it was most likely gaseous. (Trenn, 1974)

1901 (October)
Soddy and Rutherford make the first observation of natural nuclear disintegration of an unstable element. They reported the change of thorium to thorium-X. They knew that thorium-X was a gas in the argon family, but they did not make a transmutation claim, nor did they specifically claim that they had created a new element. The gas was later identified as radon. (Rutherford and Soddy, 1902, Compounds. I; 1902, Compounds. II, 1902, Radioactivity, Part I, 1902, Radioactivity, Part II, 1902, Note)

1902 (March)
Ramsay recognizes that Rutherford and Soddy's emanation appears to be a noble gas; however, he is perplexed because its radioactive nature seems to contradict the nature of an inert chemical. (Trenn, 1974)

1903 (March)
Pierre Curie and Laborde observe the intense heat given off by radium. This becomes one of the foundational discoveries that leads to the concept of nuclear energy. (Curie and Laborde, 1903, Romer, 1964)

1903 (May)
Rutherford and Soddy publish their final joint paper and establish their Law of Radioactive Change. They propose that radiation is directly tied to atomic changes, that those changes make new elements, and that such atomic changes do, in fact, take place on Earth. (Rutherford and Soddy, 1903, Radioactivity of Uranium, 1903, Comparative Study, 1903, Condensation, 1903, Radioactive Change)

1903 (July)
Ramsay and Soddy make the first observation of natural nuclear transmutation from an unstable element. They discover that an emanation of radium is helium. This supports Rutherford and Soddy's 1901 observation of a thorium-to-thorium-X change. (Ramsay and Soddy, 1903; Freund, 1904)

1904 (March)
Thomson proposes an atomic model of the nucleus surrounded by electrons. This is the first theoretical challenge to Dalton's theory of the indivisible atom. (Thomson, 1904)

1904 (February)
Rutherford publishes his book *Radio-Activity* and suggests that the alpha particle is a helium atom. In 1909, along with Thomas Royds, he proves that the alpha particle is a helium ion. (Rutherford, 1904)

1904 (May)
Frederick Soddy publishes his book *Radio-Activity: An Elementary Treatise From the Standpoint of the Disintegration Theory*. (Soddy, 1904)

1905 (July)
Clarence Skinner observes that hydrogen is occluded in metals. This is a key factor in understanding future transmutations that occur without the explicit addition of hydrogen or deuterium. (Skinner, 1905)

1905 (September)
Albert Einstein publishes his theory of special relativity: the relationship between mass and energy, $E=mc^2$. (Einstein, 1905)

1907 (May 4)
Ramsay reports man-made transmutations from stable elements: the production of neon, argon, lithium and sodium. The experiments are performed in water with a radium source to provide alpha seed particles. The claims were never confirmed. (Ramsay, 1907, Chemical Action; Cameron and Ramsay, 1907; Cameron and Ramsay 1908, Part III, 1908, Part IV; Trenn, 1974)

1908
Soddy and Thomas D. Mackenzie observe helium after they pass high-voltage electric discharges through aluminum electrodes in empty X-ray bulbs. They conclude that the helium was occluded in the bulk of the aluminum. (Soddy and Mackenzie, 1908)

1909
Rutherford and Thomas Royds prove that the alpha particle is a helium nucleus. Rutherford suggested this in 1903. (Rutherford and Royds, 1909)

1911 (March)
Rutherford speculates that atoms have a central core, a nucleus. He directs a beam of alpha particles at a thin gold foil and observes a scattering anomaly. He proposes a model for atomic structure. (Rutherford, 1911)

1912 (July)
Ramsay informally reports man-made transmutations from hydrogen that create helium and neon. Instead of using radium as a source of alphas as he did in 1907, he uses an electrical discharge with aluminum electrodes in a glass X-ray tube filled with hydrogen. His method and result is similar to that of Soddy and Mackenzie in 1908, but they did not claim that the helium was the result of a transmutation. (Ramsay, 1912)

1913 (February 6)
Ramsay, John Norman Collie and Hubert Patterson report the first set of confirmed man-made nuclear transmutations from stable elements. They report the production of helium and neon in glass X-ray tubes filled with hydrogen, and sometimes with some oxygen, subjected to an electric discharge. (Collie and Patterson, 1913, Origins, 1913, Low Pressures, 1913, Part II; Ramsay, 1913)

1913 (March)
Thomson unknowingly successfully replicates the Collie and Patterson experiments. He also reports finding X_3, which, in 1934, is identified as tritium. (Thomson, 1913, *Nature*, 1913, *Science*)

1913 (June)
James Irvine Orme Masson successfully replicates the Collie and Patterson experiments. (Masson, 1913)

1913 (July)
Niels Bohr introduces his model of the atom. He completes publication of his model in 1921. (Bohr, 1913)

1913 (December)
Soddy coins the term and concept of "isotopes" for atoms that have the same nuclear charge but different mass. (Soddy, 1913, Intra-Atomic)

1913 (December 11)
Rutherford confirms the existence of the atomic nucleus. (Rutherford, 1913, 1914, Structure)

1914 (April)
Winchester unknowingly successfully replicates the Collie and Patterson experiments. (Winchester, 1914) See also Skinner 1905.

1914 (November 2)
Collie, Patterson and Masson, in response to critique, vaporize the metals and the glassware and find no occluded helium or neon. (Collie, Patterson and Masson, 1914)

1919 (June)
Rutherford reports the first man-made disintegration of a stable element. He discovers that, when an alpha particle impacts a nitrogen atom, a proton is ejected. Separately, his speculation about the primary mechanism was incorrect. He did not, contrary to popular myth, discover or report the transmutation of nitrogen to oxygen. (Rutherford, 1919, Collisions IV) See also Patrick Maynard Stewart Blackett, 1925.

1919 (June)
Rutherford proposes that the atom contains a neutral particle. He also suggests that a proton and electron might combine to form a neutral particle. (Rutherford, 1920, Stuewer, 1986) See also 1932, James Chadwick experimentally confirmed the existence of the neutron; 1934, Gian-Carlo Wick theorized about production of neutrons by electron-capture.

1920 (August)
At the 1920 meeting of the British Association for the Advancement of Science, Rutherford suggests that the nucleus of hydrogen should be called a proton. Rutherford's more precise concept of proton replaces Prout's 1816 protyle idea and follows the naming pattern of Thomson's electron. After some discussion at the meeting, the name is accepted. (Eve, 1939)

1920
Charles Galton Darwin describes collective many-body excitations of electrons. His ideas lay the foundation for the understanding of the collective behavior of electrons. (Darwin, 1920)

1921 (March 24)
Bohr publishes his theory of electron configurations and atomic structure. He had started publishing his model in 1913. (Bohr, 1921)

1921
James Chadwick and Etienne S. Bieler expand the understanding of strong-interaction collisions between alpha particles and hydrogen nuclei. (Chadwick and Bieler, 1921)

1922 (March)
Wendt and Irion present results of a man-made nuclear transmutation from a stable element. They use the exploding-wire method, running a high-voltage discharge inside an evacuated bulb through a tungsten wire, and obtain helium. Rather than transmuting tungsten, they may have transmuted hydrogen into helium because it is almost always present as a monolayer on the surfaces of metals. (Wendt and Irion, 1922) See also Skinner 1905, Winchester 1914.

1923 (November)
Robert Millikan speculates that transmutations going from lighter to heavier elements might occur in the stars. The idea of nuclear disintegration going from heavy to light elements is known at this time. (Millikan, 1923)

1924 (July 18)
Adolf Miethe makes claim of mercury to gold, and possibly platinum transmutation in high-powered electric lamps. (Miethe, 1924)

1925 (February 2)
Patrick Maynard Stewart Blackett, a research fellow working under the direction of Rutherford, after conducting four years of experiments, reports a man-made transmutation from a stable element. He observes the transmutation of nitrogen to oxygen. He correctly proposes that the nitrogen atom captures and integrates the alpha particle, that it temporarily creates a fluorine atom, then disintegrates into a proton and oxygen atom. Rutherford is acknowledged in the paper, but he did not perform the experiments, and he is not a co-author. (Blackett, 1925)

1925 (July)
Hantaro Nagaoka reports results of a man-made nuclear transmutation from a stable element. He claims to transmute tungsten into stable gold and platinum. He assumes, apparently incorrectly, that the source material was mercury, rather than the hydrogen present in the transformer oil of his experiment. (Nagaoka, 1925)

1925 (August 7)
Arthur Smits and A. Karssen publish claim of lead to mercury and thallium transmutation. (Smits and Karssen, 1925)

1926 (September 15)
Fritz Paneth and Kurt Peters report a man-made transmutation from a stable element. They observe the transmutation of hydrogen to helium using a method of hydrogen absorption going into finely divided palladium within a red-hot glass capillary. (Paneth and Peters, 1926)

1929
Fritz Houtermans and Robert Atkinson propose that thermal kinetic energies inside stars are high enough to allow nuclei of light elements to overcome the Coulomb barrier and form heavier elements. The concept is later identified as thermonuclear fusion. In 1933, Oliphant experimentally confirms nuclear fusion. (Atkinson and Houtermans, 1929)

1932
James Chadwick experimentally confirms the existence of the neutron. In 1910, Rutherford had theorized its existence. (Chadwick, 1932)

1932
John Cockcroft and Ernest Walton, at the Cavendish laboratory, build the first apparatus for accelerating atomic particles to high energies. They report the first man-made transmutation by artificially accelerated particles: the disintegration of lithium by fast hydrogen protons. (Cockcroft and Walton, 1932)

1933
Leó Szilárd conceives the idea of the nuclear chain reaction. (Rhodes, 1986)

1933
Mark Oliphant experimentally confirms fusion of deuterons into various targets to create helium-3 and tritium, using a particle accelerator at Cavendish. Concurrently, he observes the liberation of excess nuclear binding energy, which prompts him to speculate that fusion is the process that powers the sun. (Oliphant and Rutherford, 1934)

1934 (January)
Frédéric and Irène Joliot-Curie create artificial radioactivity in previously stable elements. (Joliot and Joliot-Curie, 1934)

1934 (March 3)
Gian-Carlo Wick proposes the concept of electron capture. (Wick, 1934)

1935
Harold John Taylor reports the first neutron-capture-based transmutations. He experimentally demonstrates that boron-10 nuclei capture thermal neutrons and fission into helium-4 and lithium-7. (Taylor, 1935)

1938 (November 18)
Otto Hahn and Fritz Strassmann bombard uranium with neutrons and transmute uranium into smaller atoms. In 1939, Lise Meitner and Otto Frisch identify the effect as fission. (Hahn and Strassmann, 1938)

1939 (February 11)
Lise Meitner and Otto Frisch propose the concept of neutron-induced uranium fission. (Meitner and Frisch, 1939)

1939 (March 1)
Hans Bethe proposes that 1) hydrogen-hydrogen fusion is the process that powers the stars, 2) no elements heavier than helium-4 could have been formed in stars, and 3) the production of neutrons in stars is negligible. His first proposal was correct; the second and third were not. (Bethe, 1939)

1939 (August)
Leó Szilárd writes a letter, signed by Einstein, to President Franklin D. Roosevelt and encourages him to develop an atomic fission bomb.

1942
Roosevelt creates the Manhattan Project to develop the atomic fission bomb in response to Szilárd's letter and prevailing fears of a German atomic bomb under development.

1946 (March 14)
Fred Hoyle makes an early contribution to the theory of supernovae-exploding stars. He proposes that neutron creation in the hot cores of collapsing stars can be explained by the reaction of an electron with a proton ($e + p \rightarrow n + \nu$). (Hoyle, 1946)

1951
Ernest Sternglass, a Ph.D. candidate at Cornell University, observes neutron production in keV-energy (low-energy) electric discharge experiments in a hydrogen-filled X-ray tube directed at targets of silver and indium. See Darwin 1920. (Sternglass, 1997, See also *Hacking the Atom*, Krivit, 2016)

1957
Geoffrey and Margaret Burbidge, William Fowler and Fred Hoyle propose what is later regarded as the modern concept of nucleosynthesis of elements in stars. They theorize that fusion reactions create elements up to the atomic mass of iron and that neutron capture processes and decays create heavier elements beyond iron. Prior to their work fusion-based concepts alone were unable to fully explain the production of heavier elements The group later propose that nucleosynthesis also occurred outside the cores of stars. (Burbidge, Burbidge, Fowler and Hoyle, 1957)

Appendix F - Definition* of Low-Energy Nuclear Reactions

Low-energy nuclear reactions (LENRs) are a class of nuclear reactions — based on Standard Model physics — that can occur in condensed matter under mild macrophysical conditions. Key steps in LENR processes, unlike nuclear fusion or fission, are based primarily on electroweak interactions rather than strong-force interactions. Unlike fission reactions, low-energy nuclear reactions do not produce nuclear chain reactions.

LENRs involve a broad set of nuclear phenomena spanning many length-scales that have two characteristics in common: a) production of neutrons from electroweak reactions; and b) many-body collective effects between oppositely charged particles. (In condensed-matter systems, these particles are typically quantum mechanically entangled).

LENRs take place in three realms: a) electrically dominated reactions, in which nuclear-strength local electric fields on micron scales in condensed matter enable electroweak neutron production; b) magnetically dominated reactions, in which many-body collective magnetic-field effects directly accelerate charged particles in plasmas; and c) mixed reactions, in which components of dusty plasmas behave in ways characteristic of the electrically dominated reactions and the magnetically dominated reactions.

The word "low" in "low-energy nuclear reactions" refers to the magnitude of input energies that are required to trigger LENR reactions; the magnitude of output energies released after triggering may be either low or high. Researchers chose this term to distinguish it from the field of high-energy particle physics, which uses very high temperatures or particle accelerators to trigger nuclear reactions.

The two most unusual characteristics of LENRs are that neutron-catalyzed transmutation reactions, which typically occur only in stars, fission reactors, or high-energy particle accelerators, can be initiated in tabletop condensed-matter experimental systems without releasing biologically dangerous amounts of energetic neutron or gamma radiation.

Electrically Dominated LENR Reactions

Electrically dominated reactions take place in condensed matter. These

LENRs take place under relatively mild conditions — that is, without the requirement of using large nuclear fission reactors, extremely high temperatures, or high-energy particle accelerators.

Given proper types and amounts of input energy, these LENRs take place when specific conditions are present on the surfaces of metals or at metal-oxide interfaces, in the presence of hydrogen or ones of its isotopes, deuterium or tritium. No radioactive seed elements are required. Neutrons produced in an electroweak reaction at micron-scale LENR-active sites on surfaces or at interfaces are subsequently captured by nearby atoms; these energy-releasing captures induce nuclear transmutations. Neutrons produced in LENRs have ultra-low energy, so almost all of them are captured locally; externally detectable emissions of deadly energetic neutrons are thus also avoided.

LENR experiments typically produce a variety of nuclear transmutation products and various types of effects and may produce macroscopically measurable excess heat. A variety of elements may be synthesized from one another, and isotopic shifts may occur; these transmutation products are generally stable elements produced by beta decays of short-lived, neutron-rich unstable isotopes created by previous neutron captures.

According to the Widom-Larsen theory, LENRs in the electrically dominated realm have two unique characteristics: a) produced neutrons have ultra-low-momentum and b) unreacted heavy electrons present in LENR-active sites suppress dangerous energetic gamma emission by locally converting incident gamma radiation from any source directly into infrared radiation (heat). (Widom and Larsen, 2006; Srivastava et al. 2010)

Magnetically Dominated LENR Reactions

Magnetically dominated reactions take place in plasmas. These LENRs can, for example, occur in magnetic flux tubes of solar flares; these processes may produce GeV neutrons, other elementary particles, and energetic gamma rays that are not suppressed.

Mixed LENR Reactions

Mixed reactions take place in dusty plasmas. These LENRs can occur in organized magnetic fields present in a plasma as well as on solid surfaces of micron- to nanometer-sized dust particles of condensed matter, which are embedded in such plasmas. Examples include exploding wire experiments and in natural lightning.   * *(See also concise LENRs definition in Glossary)*

Glossary of Scientific Terms

Absorber: *See* neutron absorber.

Accelerator: *See* particle accelerator.

Activation: A process in which a non-radioactive material is subjected to nuclear radiation and becomes radioactive.

Activity: A measure of the level of radioactivity of a material. Measured by the number of spontaneous nuclear disintegrations in a specific amount of material during a specific interval.

Alchemy: Primarily a reference to ancient methods and practices intended to effect elemental or personal transformation.

Alpha (particle, emission): A Greek letter used to describe one of the first types of radioactive emissions. It is emitted during a nuclear reaction and was later identified as a helium-4 nucleus. (*See also*: Appendix A)

Alpha decay: Radioactive decay in which an alpha particle is emitted. Each emitted alpha lowers the atomic number of the nucleus by two and its atomic mass by four.

AMU: Atomic Mass Unit (See atomic mass)

Anode: In electrolysis, the metal contact point of the electrical circuit that attracts the flow of electrons.

Atom (atomic): Basic building block of all matter. Atoms comprise three elementary particles: protons, neutrons and electrons. Each atom has one nucleus in its center containing the protons and neutrons. The nucleus is surrounded by electrons, normally equal in number to the number of protons in the nucleus of a neutral atom.

Atomic energy: *See* nuclear energy.

Atomic number: Measured by the total number of protons in an atom's nucleus; determines the type of chemical element. *See also* atomic mass.

Atomic mass: Effectively measured by the total number of protons and neutrons in an atom's nucleus; determines the type of isotope within a specific range of possible isotopes. *See also* atomic number.

Atomic transformation: *See* transmutation.

Beta (particle, emission): A Greek letter used to describe one of the first types of radioactive emissions. It is emitted during a nuclear reaction and was later identified as an electron. (*See also*: Appendix A)

Beta decay: Radioactive decay in which a beta particle is emitted or in which an inner-shell orbital electron capture by the nucleus occurs.*

Binding energy: For a nucleus, the energy required to pull the neutrons and protons apart.

BF3 detector (counter): Boron tetrafluoride detector; used to measure neutron emissions. Detector consists of a cylindrical tube filled with boron trifluoride gas, which is used to detect low energy "thermal" neutrons. With the addition of a neutron moderator surrounding the detector (to bring down neutron energy), the detector can also be used to detect higher-energy "fast neutrons."

Branching ratio (D+D fusion): According to the well-understood theory of deuterium-deuterium nuclear fusion, the reaction paths occur through one of three possible branches. The first branch produces a neutron. The second branch produces tritium. The third branch produces helium-4. In D+D fusion reactions, on average, a neutron is produced almost 50% of the time, tritium is produced almost 50% of the time, and helium-4 is produced less than 1% of the time. Since the discovery of D+D fusion, these ratios have always been consistent.

Capture: *See* nuclear capture.

Cathode: In electrolysis, the metal contact point of the electrical circuit that emits the flow of electrons and attracts positive ions or protons.

Charged particle: A fundamental particle such as an electron, proton or positron, or a compound particle that carries a net positive or negative electrical charge.

Chemistry: An area of science primarily involved with interactions between atoms and electrons, their structures and properties. There are two historical exceptions. The first was in the early 20th century, when chemists were as involved in nuclear research as physicists were. The second period began in 1989, with the introduction of what was later called LENRs.

Cold fusion (idea): The proposed concept that nuclear fusion reactions occur at or near room temperature. An unproven idea that deuterium nuclei overcome the Coulomb barrier at room temperature at high reaction rates and undergo deuterium-deuterium nuclear fusion. Fusion relies primarily on strong-force interactions and normally requires temperatures in the millions of degrees.

Cold fusion (history): Historical events that took place primarily in 1989 in the aftermath of the announcement of the claim by electrochemists Martin Fleischmann and Stanley Pons at the University of Utah of a "sustained nuclear fusion reaction" at room temperature.

Collective effects: Describes the interaction of many-body groups of essentially identical items, such as elementary particles. When the items interact as a group, they create different effects than they would produce either alone or with a few others. The concept can apply to many-body physics, such as electrons oscillating together, or to a flock of birds flying in formation and thus creating lift efficiencies that none of the birds could create individually.

Coulomb barrier: An electrostatic barrier surrounding positively charged nuclei that, under normal temperatures and pressures, prevents nuclei from interacting with each other.

Cross-section: A measure of the probability of a specified interaction between an incident photon or particle radiation and a target particle or system of particles. It is the reaction rate per target particle for a specified process divided by the flux density of the incident radiation.*

Decay: *See* radioactive decay.

Deuterium: A stable isotope of hydrogen that has one proton and one neutron in its nucleus. Also known as heavy hydrogen.

Deuteron: The nucleus of a deuterium atom, comprising one proton and one neutron.

Disintegration: A process in which constituent parts of the nucleus of an atom separate from the nucleus and fly off, leaving a smaller atom in place.

Electrochemist: Person who works in the field of electrochemistry.

Electrochemistry: The study of electricity and how it relates to chemical reactions. In electrochemistry, electricity can be generated by movements of electrons from one element to another in a reaction known as redox reaction, or oxidation-reduction reaction. (U.C. Davis)

Electrode: In electrochemistry, the metal contact point of the electrical circuit that conducts the flow of electrons.

Electrolysis: Chemical decomposition by an electric current of a liquid, or solution containing ions, into constituent elements.

Electrolytic fusion: The idea of creating nuclear fusion by electrolysis.

Electromagnetic force: One of the four fundamental physics forces. Repels protons from one another and keeps atomic nuclei separate from one another. *See also* strong force.

Electron: A stable elementary particle that is a component of an atom. It possesses a negative electrical charge and exists outside and orbits around the nucleus of an atom.

Electron-volt (MeV, Mega-electron-volt): A unit of energy equal to the change in energy of one electron in passing through a voltage difference of 1 volt.*

Electroweak interaction: A term used in particle physics that describes the unified behavior of electromagnetism and weak interactions.

Element: Designates a form of matter that is distinguished by a unique number of protons in its nucleus and unique chemical properties.

Emissions: *See* alpha, beta and gamma rays.

Energy: Power during a given period.

Exploding electrical conductors (Exploding wire phenomenon): A phenomenon in which an electrical conductor such as wire, foil, or film is deliberately subjected to a very high current with near instantaneous rise time (less than 2 microseconds), causing it to explode loudly and violently, momentarily forming a plasma, and leaving behind a cloud of metal vapor.

Fission: *See* nuclear fission.

Fusion: *See* nuclear fusion.

Gamma rays (gamma radiation, gamma emission): Gamma rays are highly penetrating forms of electromagnetic radiation emitted from nuclear transitions. Gamma rays are a class of photons (a larger group of massless entities) that, according to quantum mechanics, behave both as waves and as particles. On Earth, they are encountered from radioactive material decays and a few rare terrestrial events. Gamma rays are identified by their energy from the so-called photo-peak. A range of various-energy gamma ray interaction-related peaks and continua are depicted in a typical gamma spectrum ranging from the photo-peak at the upper end of the energy scale down to zero. (*See also*: Appendix A)

Gamow factor: The probability that two nuclear particles will overcome the Coulomb barrier and undergo nuclear fusion reactions.

Gas-loading: An experimental method in which molecules of a gas, typically deuterium or hydrogen, dissociate, ionize and then move into hydride-forming sites in a host metal or metal-oxide structure.

Half-life: The time required for half of a given quantity of a radioactive material to decay.

Heavy hydrogen: *See* deuterium.

Heavy water (D_2O): Water molecules made from deuterium instead of hydrogen. *See also* light water.

Hydride (Deuteride): Compounds that hydrogen (or deuterium) form with other chemical elements, typically within metals or alloys. Some metals can absorb hundreds of times their own volume of hydrogen or deuterium.

Hydrogen: A chemical element with a single proton in its nucleus. Its normal isotope, known as protium, is stable. Its second isotope, known as deuterium, is also stable. Its third isotope, known as tritium is unstable.

Ion: An atom that has either an excess or shortage of an electron or electrons. Ordinarily, neutral atoms have an equal number of electrons to their protons.

Ionization: A process by which an atom gains or loses an electron or electrons.

Isotope: A variation of an element that contains the same number of protons but a different number of neutrons from the most abundant version of that element. Isotopes have the same atomic number but a different atomic mass.

Isotope, stable: An isotope that is not undergoing radioactive decay or emitting gamma radiation.

Isotope, unstable: An isotope that is undergoing radioactive decay involving emission of particles and/or gamma radiation.

Isotopic abundance: The relative number of atoms of a specific isotope among all the isotopes of a given element, expressed as a fraction of all the isotopes of that element.

Isotopic shift: A change in the ratios among isotopes of one species of elements away from the isotopic abundance of the same species that exists in nature.

LENR, LENRs: *See* low-energy nuclear reactions.

Light water (H_2O) (normal water): Water composed of the normal hydrogen isotope, which contains one proton and no neutron. The term "normal water" is used sometimes synonymously; however, one of every 6,000 molecules of normal water is a molecule of heavy water.

Loading: The process of placing atoms of deuterium (or hydrogen) interstitially into vacant spaces within the crystalline lattice of metallic elements.

Loading ratio: The ratio between the number of atoms of deuterium (or hydrogen) and the number of atoms of the host metal into which they have been loaded.

Low-energy nuclear reactions (LENRs): A class of nuclear reactions that occur at or near room temperature that are based on non-fusion reactions, for example, neutron-based reactions. LENRs are based on Standard Model physics and can occur in condensed matter under moderate (room-temperature) conditions. Key steps in LENR processes, unlike nuclear fusion or fission, are based primarily on electroweak interactions rather than strong-force interactions. Unlike fission reactions, low-energy nuclear reactions do not produce nuclear chain reactions. (*See also* Appendix F)

Metal hydrides or deuterides: Metals that have absorbed hydrogen or deuterium in their atomic structure, or lattice. *See also* loading.

Moderator: *See* neutron moderator.

Neutrino: An elementary particle having virtually no mass. Like a neutron, it has no electrical charge; it barely interacts with ordinary matter.

Neutron: An unstable (when outside a nucleus) elementary particle that is a component of an atom and exists inside the nucleus of an atom. It has no electrical charge. A free neutron outside of a nucleus has a half-life of approximately 10.3 minutes before it decays into a proton, an electron, and an electron antineutrino.

Neutron absorber: A material or object with which neutrons interact, resulting in their disappearance as free particles without production of other neutrons.

Neutron capture: A nuclear reaction in which an atomic nucleus and one or more neutrons collide and merge to form a heavier nucleus.

Neutron, cold: Neutrons of kinetic energy on the order of 1 milli-electron-volt or less (0.001 eV).

Neutron, fast: Neutrons having kinetic energy between 1 MeV and 20 MeV.

Neutron moderator: Material used to reduce the speed of neutrons, without absorbing them into the moderator material.

Neutron, prompt: Neutrons emitted from a nuclear process, at the time of the reaction, without measurable delay.

Neutron, slow: Neutrons having kinetic energy between 1 eV and 10 eV.

Neutron, thermal: A free neutron that has been slowed down by a moderator, is in equilibrium with its surroundings, has an energy between 0.025 eV and 0.2 eV.

Neutron, ultra-low-momentum (ULMN): A neutron with kinetic energies that are effectively zero, on the order of 10^{-12} eV or less — that is, .000000000001 eV. The kinetic energy of ULMNs is an estimated value because it has never been measured. ULMNs are extremely slow neutrons with extremely low kinetic energies and commensurately large DeBroglie wavelengths because they are created through a many-body collective process (as opposed to being produced by a two-body nuclear reaction that occurs inside a star). ULMNs are thus orders of magnitude slower than so-called "ultra-cold" neutrons, which are typically produced for experiments aiming to better measure the lifetime of free neutrons located outside nuclei. (Courtesy: Lewis Larsen)

Nuclear: Activity or properties having to do with characteristics of or changes in an atomic nucleus.

Nuclear capture: A nuclear process by which an atom acquires an additional particle.

Nuclear chemistry: Chemistry-related aspects of nuclear and atomic research.

Nuclear energy: Energy that is released during a nuclear reaction, such as nuclear fission, nuclear fusion, radioactive decay or a variety of nuclear processes that capture nuclear particles.

Nuclear fission: The process in which a larger nucleus is split into two (or, rarely, more) parts. The process is usually accompanied by the emission of neutrons, gamma radiation and, rarely, small charged nuclear fragments.

Nuclear fission, spontaneous: Nuclear fission that occurs spontaneously, without the addition of particles or energy to the nucleus.*

Nuclear fusion: The process in which two light nuclei overcome electromagnetic repulsion and form one newer, heavier atom.

Nuclear physics: Approaches to and studies of nuclear science and technology based on principles, processes and devices common to physics.

Nuclear process: A mechanism, such as fission, fusion, or radioactive decay, which changes the energy, form, or structure of the nucleus.

Nuclear reaction: An event, occurring from a nuclear process, in which the energy, form, or structure of the nucleus is changed.

Nuclear science: The study of nuclear processes and reactions.

Nuclear transformation: *See* nuclear transmutation.

Nuclear transmutation: A nuclear process in which an element changes into another element by the increase or decrease in the number of protons in its nucleus.

Nuclei: *See* nucleus.

Nucleosynthesis: The formation of new nuclides by any number of nuclear processes, including nuclear decay.

Nucleus (nuclei, pl.): Center part of an atom that contains protons and neutrons. Comprises nearly the entire mass of the atom but only a tiny part of its total volume.

Nuclide: A distinct species of an atom identified by the number of protons and neutrons in its nucleus and its nuclear energy state.

Particles: *See* charged particle and neutron.

Particle accelerator: A device for imparting kinetic energy to charged particles.*

Photon: A massless elementary particle that is a unit of light and other forms of electromagnetic radiation. It can have properties of both waves and particles.

Physicist, nuclear: Person who works in the field of nuclear physics.

Physics: A field of science and technology that measures, studies and influences matter, motion and energy.

Power: The rate of doing work; equivalent to the amount of energy consumed per unit of time.

Proton: Stable elementary particle that is a component of atoms. The proton has a positive electrical charge and exists in the nucleus of an atom.

Quantum mechanics: The branch of mechanics that deals with the mathematical description of the motion and interaction of subatomic particles, incorporating the concepts of quantization of energy, wave-particle duality, the uncertainty principle, and the correspondence principle. (Source: Oxford Dictionary)

Radiation, nuclear (Radioactive Emission): Emission of charged or uncharged particles or electromagnetic rays, including alphas, betas, neutrons, and gamma-rays.

Radiation, prompt: Prompt radiation is produced and emitted from its source immediately. When the reaction stops, so does the prompt radiation. *See also* radioactive decay.

GLOSSARY • 345

Radioactive: The property of an unstable material that spontaneously emits particles or gamma rays.
Radioactive decay: A form of nuclear radiation that emits alpha and/or beta particles from radioactive materials. The emissions may take place during nuclear reactions as well as after the reactions stop. The decay causes the radioactive interior to lose some of its constituent material. An element that undergoes radioactive decay will change into a new element or a new isotope.
Radioactive half-life: For a single radioactive decay process, the time required for the activity to decrease to half its value by that process.*
Radioactivity: A naturally occurring process in which unstable elements spontaneously emit particles or gamma rays. In addition to naturally occurring radiation, man-made nuclear processes can cause some non-radioactive elements to become radioactive.
Radiochemist: Person who works in the field of radiochemistry.
Radiochemistry: The part of chemistry that deals with radioactive materials.*
Radioisotope, *See* Isotope, unstable
Radionuclide: A radioactive nuclide.
Radium: An unstable radioactive chemical element with a half-life of about 1,600 years, and 88 protons in its nucleus.
Radon: An unstable chemical element and radioactive gas with a half-life of 3.8 days, and 86 protons in its nucleus. It is produced from the decay of uranium or thorium.
Scattering: A process in which a change in direction or energy of an incident particle or incident radiation is caused by a collision with a particle or a system of particles.*
Scattering, elastic: Scattering in which the total kinetic energy is unchanged.*
Sonic implantation: An experimental method that uses acoustic cavitation to stimulate activity on metal surfaces and induce low-energy nuclear reactions.
Spectral lines: Bright and dark lines — seen in spectra of photon-emitting items, such as candle flames, glowing gas, or stars — that are characteristic of a given atom or molecule.
Standard Model: The Standard Model of particle physics is a theory that explains the physics of the world and what holds it together. It encompasses the behavior of fundamental particles across three of the four fundamental forces in physics: electromagnetic, weak interactions, and strong interactions (not gravity).
Strong force: One of the four fundamental physics forces; works only at very short distances within nuclei. Keeps protons and neutrons bound together inside atomic nuclei. *See also* electromagnetic force.
Surface plasmon electrons: A collective many-body effect; coherent oscillations of entangled electrons that take place at the surface of metals and at other interfaces.
Transformation: *See* transmutation.
Transmutation, natural: Spontaneous nuclear transmutation that occurs by the natural activity of a radioactive element.

Transmutation, nuclear: The changing of one element to another by a change in the number of protons in its nucleus.

Transmutation, man-made: Human-triggered nuclear transmutation; traditionally occurs by exposure to radioactive sources. Can also occur by nontraditional processes, specifically LENRs.

Tritium: An unstable isotope of hydrogen. A chemical element with a single proton in its nucleus and two neutrons. It is radioactive with a half-life of 12.3 years.

Weak force: A fundamental force of physics that produces weak interactions.

Weak interaction: An elementary particle interaction that is involved in many forms of nuclear decay (radioactivity), for example a beta decay process. In all such interactions, neutrinos are emitted or absorbed. Weak interactions are distinct from strong-force interactions because, at low average particle energies, weak-interaction cross sections are vastly lower than strong-force interactions. Weak interactions are not necessarily weak energetically, and some can involve very large releases of energy. For example, beta decays of some extremely neutron-rich nitrogen isotopes can release more than 20 MeV of nuclear binding energy. For comparison, the strong-force deuterium-tritium fusion reaction releases 17.6 MeV.

* Source: Glossary of Terms in Nuclear Science and Technology, American Nuclear Society, ISBN 0894485539

Bibliography

Adrian Daily Telegram (March 30, 1925) "Scientists Believed About to Transmute Base Metal to Gold," p. 7

Allison, Samuel King and Harkins, William Draper (April 1924) "The Absence of Helium From the Gases Left After the Passage of Electrical Discharges: I, Between Fine Wires in a Vacuum; II, Through Hydrogen; and III, Through Mercury Vapor," *Journal of the American Chemical Society*, **46**(4), p. 814-24

Atkinson, R. and Houtermans F. (1929) "Aufbaumöglichkeit in Sternen," *Z. für Physik*, **54**, p. 656-665

Badash, Lawrence (1966) "How the 'Newer Alchemy' Was Received," *Scientific American*, **213**:88-95

Baily, Charles (Oct. 12, 2008) "Atomic Modeling in the Early 20th Century: 1904-1913," Contributed Talk, 24th Regional Conference on the History and Philosophy of Science, Boulder, Colo.

Balenovic, Ivan, transl. and adapt. (unpublished, 2007) *Milorad Mladjenovic: An Autobiography* (Novi Sad, Serbia)

Baly, E.C.C. and Riding, R.W. (Oct. 30, 1926) "The Occurrence of Helium and Neon in Vacuum Tubes," *Nature*, **118**(2974), p. 625-26

Becquerel, Alexandre Edmond (1868) "*La Lumière: Ses Causes et Ses Effets, Vol. 2*," Librairie de Firmin Didot

Becquerel, Henri (1896) "On the Invisible Rays Emitted by Phosphorescent Bodies," read before the French Academy of Science, March 2, 1896, *Comptes Rendus* **122**, p. 501

Bethe, Hans, (March 1939) "Energy Production in Stars," *Phys. Rev.*, **55**, p. 434-456

Blackett, Patrick Maynard Stewart (Feb. 2, 1925) "The Ejection of Protons From Nitrogen Nuclei, Photographed by the Wilson Method," *Journal of the Chemical Society Transactions*. Series A, **107**(742), p. 349-60

Bohr, Niels (July 1913) "On the Constitution of Atoms and Molecules," *Philosophical Magazine*, Series 6, **26**, p. 1-25

Bohr, Niels (March 24, 1921) "Atomic Structure," *Nature*, **107**, p. 104-7

Bolton, H. Carrington (Oct. 17, 1897) "The Revival of Alchemy," *New York Times*, p. 6

Bolton, H. Carrington (Dec. 10, 1897) "The Revival of Alchemy," *Science*, p. 853-63

Boston Globe (May 30, 1920) "Turning Lead Into Gold"

Brandon Daily Sun (April 13, 1925) "Close to Success in Age-Old Aim of Alchemy"

Briscoe, Henry Vincent Aird, Robinson, Percy Lucock, and Stephenson, George Edward (1925) "The Electrical Explosion of Tungsten Wires," *Journal of the Chemical Society Transactions*, **127**, p. 240-47

Burbidge, E. Margaret, Burbidge, Geoffrey R., Fowler, William A. and Hoyle, Fred (1957) "Synthesis of the Elements in the Stars," *Rev. Mod. Phys.* **29**(4), p. 547–650

Bureau of Standards (Dec. 21, 1922) "Bibliography of Scientific Literature Relating to Helium," No. 81

Cameron, Alexander Thomas and Ramsay, William (1907) "The Chemical Action of the Radium Emanation: Part II. On Solutions Containing Copper, and Lead, and on Water," *Journal of the Chemical Society Transactions*, 91, p. 1593-606

Cameron, Alexander Thomas and Ramsay, William (1908) "The Chemical Action of the Radium Emanation: Part III. On Water and Certain Gases," *Journal of the Chemical Society Transactions*, 93, p. 966-92

Cameron, Alexander Thomas and Ramsay, William (1908) "The Chemical Action of the Radium Emanation: Part IV. On Water," *Journal of the Chemical Society Transactions*, 93, 992-7

Campbell, John (November 10, 1999) "Rutherford - Scientist Supreme," *AAS Publications Canton Times*

Canton Times (April 2, 1920) "The Beginnings of Transmutation?", p. 3

Chace, William, G, (May 16, 1963) "Exploding Wires and Their Uses, *New Scientist*, p. 366-8

Chace, William G. and Watson, Eleanor M. (1967) *A Bibliography of the Electrically Exploded Conductor Phenomenon*, 4th ed., Armed Services Technical Information Agency

Chadwick, James and Bieler, Etienne S. (1921) "Collisions of Alpha Particles With Hydrogen Nuclei," *Philosophical Magazine*, Series 6, 42, p. 923

Chadwick, James (1932) "Possible Existence of a Neutron," *Nature*, 129, p. 312; Chadwick, James (1932) "The Existence of a Neutron," *Proc. Roy. Soc.*, A136, p. 692

Charleston Daily Mail, (Nov. 23, 1924) "Cheap Way to Turn Mercury Into Gold Scientist's Aim," p. 8

Cockcroft, John D. and Walton, Ernest T. S. (1932) "The Disintegration of Elements by High-Velocity Protons," *Proceedings of the Royal Society of London,* 137, p. 229; Cockcroft and Walton (1932) *Nature* 129, p. 242

Collie, John N. (Aug. 1, 1914) "Note on the Paper by T. R. Merton on 'Attempts to Produce the Rare Gases by Electric Discharge,'" *Proceedings of the Royal Society of London*, 90-A (621), p. 554-6

Collie, John N., and Patterson, Hubert S. (1913) "The Presence of Neon in Hydrogen After the Passage of the Electric Discharge Through the Latter at Low Pressures," *Journal of the Chemical Society Transactions*, 103, p. 419-26

Collie, John N., and Patterson, Hubert S. (Feb. 13, 1913) "Origins of Helium and Neon," *Nature*, 90, p. 653-4

Collie, John N., and Patterson, Hubert S. (June 28, 1913) "The Presence of Neon in Hydrogen After the Passage of the Electric Discharge Through the Latter at Low Pressures: Part II," *Proceedings of the Chemical Society, London*, 29(418) 217-21

Collie, John N., Patterson, Hubert S., and Masson, Masson, James Irvine Orme (Nov. 2, 1914) "The Production of Neon and Helium by the Electrical Discharge," *Proceedings of the Royal Society of London*, 91-A (623), p. 30-45

Curie, Marie (July, 1898) "Rayons Émis Par Les Composes De L' Uranium et du Thorium," *Comptes Rendus de l' Académie des Sciences Paris*, 126, p. 1101-3

Curie, Marie Sklodowska (January 1904) "Radium and Radioactivity," *Century Magazine*
Curie, Marie (1963) *Pierre Curie*, Courier Dover Publications, New York
Curie, Marie and Gleditsch, Ellen (Sept. 25, 1908) "Action of Radium Emanation on Solutions of Copper Salts," *Chemical News*, p. 157-8, English translation of Curie, Marie et Gleditsch, Ellen, "Action de l'émanation du radium sur les solutions des sels de cuivre," *Comptes Rendus Hebdomadaires des Séances de l'Académie des Sciences*, Vol. CXCLVII, p. 345, (1908)
Curie, Pierre, Curie, Marie Sklodowska, and Bémont, Gustave (1898) "On a New, Strongly Radioactive Substance, Contained in Pitchblende," *Comptes Rendus*, **127**, p. 1215-17 (1966); translated and reprinted in Boorse, Henry A. and Motz, Lloyd, eds., *The World of the Atom*, **1**, Basic Books
Curie, Pierre and Laborde, Albert (1903) "On a Heat Spontaneously Released by the Salts of Radium," *Comptes Rendus de l'Academie des Science, Paris*, **136**, p. 673-5

Dahl, Per Fridtjof (2002) *From Nuclear Transmutation to Nuclear Fission, 1932-1939*, Institute of Physics Publishing, ISBN 0750308656
Daintith, John (August 18, 2008) *Biographical Encyclopedia of Scientists*, 3rd ed., Taylor & Francis
Dalton, John (May 1808) *New System of Chemical Philosophy*, R. Bickerstaff
Darwin, Charles Galton (1920) "Motion of Charged Particles," *Philosophical Magazine*, Series 6, **39**, p. 537-51
Dash, Vaidya Bhagwan (Jan 1, 1997) "Alchemy in India," in *Encyclopaedia of the History of Science, Technology, and Medicine in Non-Western Cultures*, Selin, Helaine, ed. Springer
Débarbat, Suzanne, Lequeux, James and Orchiston, Wayne (2007) "Highlighting the History of French Radio Astronomy: One: Nordmann's Attempt to Observe Solar Radio Emission in 1901," *Journal of Astronomical History and Heritage*, **10**(1), p. 3-10

Egan, Mortimer A. (May 1922) "What Won't They Do Next?," *Illustrated World*, **XXXVII**, N.3, p. 351-2
Egerton, Alfred Charles Glyn (March 1, 1915) "The Analysis of Gases After Passage of Electric Discharge," *Proceedings of the Royal Society*. Series A. **91**: A627, p.180-9
Electrician (Sept. 18, 1903) "The British Association at Southport," p. 880
Einstein, Albert (Sept. 1905) "Zur Elektrodynamik Bewegter Körper," *Annalen der Physik*, **17**(10), p. 891-921
Einstein, Albert (1923) "On the Electrodynamics of Moving Bodies," translated by George Barker Jeffery and Wilfrid Perrett, in *The Principle of Relativity*, Methuen and Company, Ltd., London
Emporia Daily Gazette (March 26, 1925) "Machine Divides Mercury," p. 9
English Mechanic and World of Science (Sept. 19, 1919) "The British Association," **2842**, p. 90
Eve, Arthur (1939) *Rutherford - Being the Life and Letters of the Rt. Hon. Lord Rutherford, OM*, Cambridge University Press

Fernandez, Bernard and Ripka, Georges (2012) *Unraveling the Mystery of the Atomic Nucleus: A Sixty-Year Journey, 1896-1956*, Springer

Fontani, Marco, Costa, Mariagrazia and Orna, Mary Virginia (2014) *The Lost Elements: The Periodic Table's Shadow Side*, Oxford University Press; 1st edition

Foote, Paul, D. (September 1924) "The Alchemist," *Scientific Monthly*, **19**(3), p. 239-62

Fouquet, Roger (March 2010) "The Slow Search for Solutions: Lessons From Historical Energy Transitions by Sector and Service," BC3 Working Paper Series, Basque Centre for Climate Change

Freund, Ida (1904) *The Study of Chemical Composition. An Account of Its Method and Historical Development, With Illustrative Quotations*, Cambridge University Press

Früngel, Frank (1965) *High-Speed Pulse Technology: Volume 1. Capacitor Discharges — Magnetohydrodynamics — X-Rays — Ultrasonics*, Academic Press, New York

Galison, Peter (1997) *Image and Logic: A Material Culture of Microphysics*, University of Chicago Press

Garrett, M.W. (Sept. 1, 1926) "Experiments Upon the Reported Transmutation of Mercury Into Gold," *Proceedings of the Royal Society of London. Series A*, **112**(761) p. 391-406

Gastonia Daily Gazette (Aug. 21, 1925) "German Scientists Find Gold From Mercury Easy," p. 1

General Electric Co. (1948) *Adventures Inside the Atom*, General Comics, sponsored by the U.S. Atomic Energy Commission

Gerward, Leif (Dec. 1899) "Paul Villard and His Discovery of Gamma Rays," *Physics in Perspective*, **1**(4), p. 367-83

Gibbs, Philip (April 8, 2010) "'Crackpots' Who Were Right 6: Abel Niépce De Saint-Victor," https://vixra.wordpress.com

Goldfein, Solomon (1978) "Energy Development from Elemental Transmutation in Biological Systems," U.S. Army Material Technology Laboratory, www.dtic.mil

Habashi, Fathi (2001) "Niépce De Saint-Victor and the Discovery of Radioactivity," *Bulletin for the History of Chemistry*, **26**(2)

Haber, Fritz (May 7, 1926) "Über den Stand der Frage nach der Umwandelbarkeit der Chemischen Elemente," *Naturwissenschaften*, **14**(19), p. 405-12

Hahn, Otto. and Strassmann, Fritz (18 Nov. 1938) "Über die Entstehung von Radiumisotopen aus Uran durch Bestrahlen mit Schnellen und Verlangsamten Neutronen," *Naturwissenschaften*, **26**(46), p. 755-756

Hansen, Steve P. (1993) *Exploding Wires: Principles, Apparatus, and Experiments*, www.belljar.net

Harkins, William D. and Shadduck, Hugh A. (Dec. 15, 1926) "The Synthesis and Disintegration of Atoms as Revealed by the Photography of Wilson Cloud Tracks," *Proceedings of the National Academy of Sciences of the United States of America*, **12**(12), p. 707-14

Harrow, Benjamin (Aug. 1919) "William Ramsay," *Scientific Monthly*, **9**(2), p. 167-78

Hasselberg, K.B. (1966) "Award Ceremony Speech for the Nobel Prize in Chemistry 1908," Dec. 10, 1908, *Nobel Lectures - Chemistry : 1901-1921*, Elsevier, p. 125-128

Heilbron, J.L., ed. (June 3, 2005) *Oxford Guide to the History of Physics and Astronomy*, Oxford University Press

Hirshberg, Leonard Keene (April 12, 1913) "Transmutation Explained Away," *Harper's Weekly*, **57**(2938), p. 21

Hoffman, Nate (1995) *Dialogue on Chemically Induced Nuclear Effects: A Guide for the Perplexed About Cold Fusion*, American Nuclear Society

Hönigschmid, Otto and Zintl, Eduard (July 17, 1925) "Über das Atomgewicht des von Miethe und Stammreich aus Quecksilber Gewonnenen Goldes," *Naturwissenschaften*, **13**(29), p. 644

Howorth, Muriel (1958) *Pioneer Research on the Atom: Rutherford and Soddy in a Glorious Chapter of Science - The Life Story of Frederick Soddy, M.A., LL.D., F.R.S., Nobel Laureate*, New World Publications

Hoyle, Fred (1946) "The Synthesis of the Elements from Hydrogen," *The Monthly Notices of the Royal Astronomical Society*, **106**(5), p. 343-383

Joliot, Frédéric and Joliot-Curie, Irene (1934) "Artificial Production of a New Kind of Radioelement," *Nature* **133**, p. 201

Journal of the Mineralogical Society of America (1925) "Notes and News," p. 76

Kansas City Star (Oct. 20, 1925) "Nothing to Replace Gold," p.3

Kansas City Star (June 27, 1926) "Claims to Have Changed Gold Into Mercury"

Kingston Gleaner (July 14, 1925) "Extract Gold From Quicksilver"

Kovarik, A.F. and McKeehan, L.W. (1926) "Radioactivity," *Bulletin of the National Research Council*, National Research Council, National Academy of Sciences, **10**(1) No. 51

L'Annunziata, Michael F. (Sept. 6, 2007) *Radioactivity: Introduction and History*, Elsevier Science

Larsen, Lewis (May 19, 2012) "LENR Transmutation Networks Can Produce Gold," Slideshare

Larsen, Lewis (Dec. 27, 2013) "Mystery of Nagaoka's 1920s Gold Experiments - Why Did Work Stop by 1930"

Masson, Masson, James Irvine Orme (June 28, 1913) "The Occurrence of Neon in Vacuum-Tubes Containing Hydrogen," *Proceedings of the Chemical Society, London*, **29**(418), p. 233

Matin (Dec. 8, 1919) "Une Immense Découverte," p.1

McGrath, James R. (May 1966) "Exploding-Wire Research 1774-1963," NRL Memorandum Report 1698, Naval Research Laboratory

Meitner, Lise and Frisch, Otto (Feb. 11 1939) "Disintegration of Uranium by Neutrons: a New Type of Nuclear Reaction," *Nature*, **143**, p. 239-240

Mellor, Joseph William, ed. (1922) *Comprehensive Treatise on Inorganic and*

Theoretical Chemistry, Vol. 1: H, O, Longmans, Green and Co.

Mellor, Joseph William, ed. (1927) *Comprehensive Treatise on Inorganic and Theoretical Chemistry*, Vol. 7, Longmans, Green and Co.

Mellor, Joseph William, ed. (1945 ed., 1st ed. 1923) *Comprehensive Treatise on Inorganic and Theoretical Chemistry, Vol. 4: Ra and Ac Families, Be, Mg, Zn, Cd, Hg*, Longmans, Green and Co.

Merton, Thomas R. (Aug. 1, 1914) "Attempts to Produce the Rare Gases by Electric Discharge," *Proceedings of the Royal Society of London* 90-A (621), p. 549-53

Miethe, Adolf (July 18, 1924) "Der Zerfall des Quecksilberatoms [The Decay of the Mercury Atom]," *Naturwissenschaften*, 12(29), p. 597-8

Miethe, Adolf (July 17, 1925) "Gold aus Quecksilber [Gold From Mercury]," *Naturwissenschaften*, 13(29), p. 635-6

Millikan, Robert A. (Nov. 1923) "Gulliver's Travels in Science," *Scribner's Magazine*, 74(5)

Millikan, Robert A. (1930) *Science and the New Civilization* [1st ed.], Charles Scribner's and Sons

Mining and Scientific Press (Feb. 11, 1922) "Synthetic Gold"

Mladjenovic, Milorad (Oct. 1992) *The History of Early Nuclear Physics, 1896-1931*, World Scientific

Montreal Gazette (June 20, 1925) "Shows Transmuted Gold in New York," p. 10

Morning Post (March 8, 1913) "The Birth of the Atom," p. 154-5

Morning Times (July 29, 1897) "Making Gold of Silver," p. 2

Nagaoka, H., Sugiura, Y., and Mishima, T. (March 29, 1924) "Isotopes of Mercury and Bismuth Revealed in the Satellites of Their Spectral Lines," *Nature*, 113(2839), p. 459-60

Nagaoka, Hantaro (Aug. 16, 1924) "Isotopes of Mercury and Bismuth and the Satellites of Their Spectral Lines," *Nature*, 114(2859), p. 245

Nagaoka, Hantaro (July 18, 1925) "Preliminary Note on the Transmutation of Mercury Into Gold," *Nature*, 116(2907), p. 95-6

National Academy of Sciences, National Research Council (1966) *Nuclear Chemistry: A Current Review*, Publication 1292-C

Nature (Aug. 29, 1907) "Mathematics and Physics at the British Association," 76(1974), p. 457-62

Nature (Feb. 13, 1913) "Origins of Helium and Neon," 90(2259), p. 653-4

Nature (Aug. 9, 1924) "The Reported Transmutation of Mercury Into Gold," 114(2858), p. 197-8

Nature (May 29, 1926) "The Present Position of the Transmutation Controversy," 117(2952)

Nature (Sept. 25, 1926) "News and Views," 118(2969), p. 455-6

Nature (Oct. 9, 1926) "The Reported Conversion of Hydrogen Into Helium," [Summary of Berichte paper], 118(2971) p. 526-7

New York Times (1880) "Alchemy and Chemistry," p. 2,

New York Times (March 26, 1909) "Transmuted Into Carbon," p. 1, 4

New York Times (Feb. 19, 1911) "Alchemy, Long Scoffed At, Turns Out to Be True; Transmutation of Metals, the Principle of the Philosopher's Stone, Accomplished in the Twentieth Century," p. SM12

New York Times (Sept. 19, 1911) "England Without Coal," p. 12
New York Times (July 21, 1912) "Helium's Power on Atoms"
New York Times (Feb. 11, 1913) "Imagination Useful in Science"
New York Times (Feb. 16, 1913) "Notes and Gleanings"
New York Times (Jan. 8, 1922) "Way to Transmute Elements Is Found"
New York Times (Jan. 25, 1922) "Gold Maker a Fake, Prof. Fisher Finds"
New York Times (March 12, 1922) "50,000 Degrees Heat Decomposes Metal," p. 22
New York Times (Nov. 23, 1924) "Synthetic Gold Might Disrupt World," p. 23
New York Times (Aug. 22, 1925) "Gold Makers Say Success Is Growing"
New York Times (Oct. 20, 1925) "Gold From Mercury Found Impossible"
New York Times (Oct. 25, 1925) "Discredit Gold by Mercury Process"
New York Times (March 5, 1926) "Miethe in New Tests Fails to Make Gold"
New York Times (Feb. 16, 1960) "Dr. Winchester, Physicist, Is Dead"
New York Tribune (Dec. 31, 1921) "Synthetic' Gold," p. 8
Nobelprize.org (retrieved March 23, 2014) "Ernest Rutherford Biography"

Oakland Tribune (Oct. 19, 1924) "The New German Gold Maker"
Ojasoo, Tiiu (Dec. 1996) "Did Becquerel Discover Radioactivity?," *Science Tribune*
Olean Evening Herald (Jan. 21, 1922) "Synthetic Gold"
Oliphant, Mark, and Rutherford, Ernest (1933) "Transmutation of Elements by Protons," *Proc. Roy. Soc. A*, **141**, p. 259; Oliphant, Mark, Harteck, P., and Rutherford, Ernest (1934) "Transmutation Effects Observed with Heavy Hydrogen," *Proceedings of the Royal Society A*, **144**, p. 692-703
Oxford Index, a Web site of the Oxford University Press (retrieved March 26, 2014) "Patrick Maynard Stewart Blackett"

Paneth, Fritz (Oct. 8, 1926) "Ancient and Modern Alchemy," *Science*, **64**(1661), p. 409-17
Paneth, Fritz (April 22, 1927) "Neuere Versuche über die Verwandlung von Wasserstoff in Helium," *Zuschriften*, **16**, p. 379
Paneth, Fritz (May 14, 1927) "The Transmutation of Hydrogen Into Helium," *Nature*, **119**(3002), p. 706-7
Paneth, Fritz (1964) *"Chemistry and Beyond: A Selection of the writings of the late Professor F.A. Paneth,"* Herbert Dingle, ed., Wiley and Sons
Paneth, Fritz, and Loleit, H. (Dec. 14, 1935) "Chemical Detection of Artificial Transmutation of Elements," *Nature*, **136**(3450), p. 950
Paneth, Fritz and Peters, Kurt (Sept, 15, 1926) "Uber die Verwandlung von Wasserstoff in Helium [The Transmutation of Hydrogen Into Helium]," *Berichte der Deutschen Chemischen Gesellschaft*, **59**(8), p. 2,039-48
Paneth, Fritz and Peters, Kurt (Oct. 8, 1926) "Ancient and Modern Alchemy," *Science*, **64**(1661), p. 409-17
Paneth, Fritz and Peters, Kurt (Oct. 22, 1926) "Über die Verwandlung von Wasserstoff in Helium [The Transmutation of Hydrogen Into Helium]," *Naturwissenschaften*, **14**(43), p. 956–62
Paneth, Fritz, Peters, Kurt and Günther, Paul (Feb. 9, 1927) "Über die Verwandlung von Wasserstoff in Helium," *Berichte der Deutschen Chemischen Gesellschaft*, **60**, p. 808-9

Perrin, Jean-Baptiste (1921) "La Structure De L'Atome," *Proceedings of Conseil de Physique Solvay Tenu a Bruxelles*, p. 68
Place, Robert Michael (Sept. 2009) *Magic and Alchemy: Mysteries, Legends, and Unexplained Phenomena*, Chelsea House
Popular Science Monthly (Aug. 1922) "Science, in Newest Feat, Explodes Atom"

Ramsay, William (1895) "Helium, a Gaseous Constituent of Certain Minerals, Part I," *Proceedings of the Royal Society of London*, **58**, p. 80-9
Ramsay, William (1907) "The Chemical Action of the Radium Emanation. Part I. Action on Distilled Water," *Journal of the Chemical Society Transactions*, **91**, p. 931-42
Ramsay, William (July 18, 1907) "Radium Emanation," *Nature*, p. 269
Ramsay, William (July 18, 1912) "Experiments with Kathode Rays," *Nature*, **89**(2229), p. 502
Ramsay, William (February 1913) "The Presence of He in the Gas From the Interior of an X-Ray Bulb," *Journal of the Chemical Society Transactions*, **103**(410), p. 264-6
Ramsay, William and Soddy, Frederick (July 28, 1903) "Experiments in Radioactivity, and the Production of Helium From Radium," *Proceedings of the Royal Society of London*, **72**, p. 204-7; reprinted in *Nature*, **68**, p. 354-5, (Aug. 13, 1903)
Redgrove, H. Stanley (1911-1922) *Alchemy: Ancient and Modern*, William Rider & Son, Ltd.; republished 2014
Reeves, Richard (Dec. 17, 2008) *A Force of Nature: The Frontier Genius of Ernest Rutherford (Great Discoveries)*, W.W. Norton
Rhodes, Richard (1986) *The Making of the Atomic Bomb*, Simon and Schuster, New York, ISBN 0671441337
Richardson, W.H. (Nov., 1958) *Exploding-Wire Phenomena*, SCR-53, Sandia Corp.
Roentgen, Wilhelm (1896) "On a New Kind of Rays," read before the Würzburg Physical and Medical Society, 1895 (translated by Arthur Stanton, *Nature*, **53**, p. 274
Romer, A., ed. (1964) *The Discovery of Radioactivity and Transmutation*, (English translation of Curie, Laborde, 1903), Dover Publications, Inc., New York
Runge, Carl (May 31, 1924) "Isotopes of Mercury and Bismuth and the Satellites of Their Spectral Lines," *Nature*, **113**(2848), p. 781
Rutherford, Ernest (1899) "Uranium Radiation and the Electrical Conduction Produced By It," *Philosophical Magazine*, Series 5, **47**, p. 109-63
Rutherford, Ernest (Jan., 1900) "A Radioactive Substance Emitted From Thorium Compounds," *Philosophical Magazine*, Series 5, **49**, p. 1-14
Rutherford, Ernest (1903) "The Magnetic and Electric Deviation of the Easily Absorbed Rays From Radium," *Philosophical Magazine*, Series 6, **5**, p. 177-87
Rutherford, Ernest (1904) *Radio-Activity*, [1st ed.], Cambridge University Press, p. 331
Rutherford, Ernest (1911) "The Scattering of Alpha and Beta Particles by Matter and the Structure of the Atom," *Philosophical Magazine*, Series 6, **21**, p. 669
Rutherford, Ernest (1913) *Radioactive Substances and Their Radiations*, Cambridge University Press

Rutherford, Ernest (Dec. 1913) "The Structure of the Atom," *Nature*, **92**, p. 423
Rutherford, Ernest (Feb. 1914) "The Structure of the Atom," *Philosophical Magazine*, Series 6, **27**, p. 488
Rutherford, Ernest (June 1919) "Collisions of Alpha Particles With Light Atoms: I. Hydrogen," *Philosophical Magazine*, Series 6, **37**, p. 537-61
Rutherford, Ernest (June 1919) "Collisions of Alpha Particles With Light Atoms: II. Velocity of the Hydrogen Atom," *Philosophical Magazine*, Series 6, **37**, p. 562-71
Rutherford, Ernest (June 1919) "Collisions of Alpha Particles With Light Atoms: III. Nitrogen and Oxygen Atoms," *Philosophical Magazine*, Series 6, **37**, p. 571-80
Rutherford, Ernest (June 1919) "Collisions of Alpha Particles With Light Atoms: IV. An Anomalous Effect in Nitrogen," *Philosophical Magazine*, Series 6, **37**, p. 581-87
Rutherford, Ernest (April 1, 1922) "Disintegration of Elements," *Nature*, **109**(2735), p. 418; (April 21, 1922), *Science*, **55**(1425), p. 422-3
Rutherford, Ernest (May 6, 1922) "Artificial Disintegration of the Elements: [Part 1]," *Nature*, **109**(2740), p. 584-6
Rutherford, Ernest (1925) "Studies of Atomic Nuclei," in *Proceedings of the Royal Institution Library of Science*, **9**, p. 73-6
Rutherford, Ernest (April 4, 1925) "Disintegration of Atomic Nuclei," *Nature*, **115**, p. 493-4
Rutherford, Ernest (1926) "The Energy in the Atom: Can Man Utilize It?," (July 1, 1924) in *Popular Research Narratives*, Vol. 2, *Fifty Brief Stories of Research, Invention or Discovery*, Williams and Wilkins, eds., Baltimore, Maryland
Rutherford, Ernest (1937) "The Newer Alchemy Based on the Henry Sidgwick Memorial Lecture, November 1936," Cambridge University Press
Rutherford, Ernest (1965) "Nuclear Constitution of Atoms," [Bakerian Lecture], *Proceedings of the Royal Society of London* **97-A**, 374-400, (1920); reprinted in *The Collected Papers of Lord Rutherford of Nelson O.M., F.R.S.* Published under the Scientific Direction of Sir James Chadwick, F.R.S., Chadwick, James, ed., **3**, George Allen and Unwin, p. 14-38
Rutherford, Ernest and Chadwick, James (March 29, 1924) "The Bombardment of Elements by Alpha Particles," *Nature*, **113**, p. 457
Rutherford, Ernest, Chadwick, James, and Ellis, Charles Drummond (1930) *Radiations From Radioactive Substances*, Cambridge University Press, p. 315
Rutherford, Ernest and Royds, Thomas (1908) "The Action of the Radium Emanation Upon Water," *Philosophical Magazine*, Series 6, **16**, p. 812-18
Rutherford, Ernest and Royds, Thomas (1909) "The Nature of the Alpha Particle From Radioactive Substances," *Philosophical Magazine*, Series 6, **17**, p. 281-6
Rutherford, Ernest and Soddy, Frederick (1902) "Note on the Condensation Points of the Thorium and Radium Emanations," *Proceedings of the Chemical Society*, p. 219-20
Rutherford, Ernest and Soddy, Frederick (April, 1902) "The Radioactivity of Thorium Compounds: I. An Investigation of the Radioactive Emanation," *Journal of the Chemical Society Transactions*, **81**, p. 321-50
Rutherford, Ernest and Soddy, Frederick (July, 1902) "The Radioactivity of Thorium Compounds: II. The Cause and Nature of Radioactivity," *Journal of*

the *Chemical Society Transactions*, **81**, p. 837-60
Rutherford, Ernest and Soddy, Frederick (Sept., 1902) "The Cause and Nature of Radioactivity, Part I," *Philosophical Magazine*, Series 4, **21**, p. 370-96
Rutherford, Ernest and Soddy, Frederick (Nov., 1902) "The Cause and Nature of Radioactivity, Part II," *Philosophical Magazine*, Series 4, **23**, p. 569-85
Rutherford, Ernest and Soddy, Frederick (April, 1903) "A Comparative Study of the Radioactivity of Radium and Thorium," *Philosophical Magazine*, Series 6, **5**, No. 28, p. 445-57
Rutherford, Ernest, and Soddy, Frederick (April, 1903) "The Radioactivity of Uranium," *Philosophical Magazine*, Series 6, **5**, No. 28, p. 441-5
Rutherford, Ernest and Soddy, Frederick (May, 1903) "Condensation of the Radioactive Emanations," *Philosophical Magazine*, Series 6, **5**(29), p. 561-76
Rutherford, Ernest and Soddy, Frederick (May, 1903) "Radioactive Change," *Philosophical Magazine*, Series 6, **5**, No. 29, p. 576-91

San Jose Evening News (Feb. 27, 1926) "Alchemy Trials Are Failure Is Word of Science"
Science Abstracts: Physics (1925) W.R. Cooper, ed., **28**, p. 392
Science Service (Dec. 19, 1924) "Gold From Mercury," *Berkeley Daily Gazette*
Science Service (Feb. 16, 1925) "Chats on Science," *Berkeley Daily Gazette*
Science Service (April 29, 1925) "Japanese Chemist Tells How He Made Gold From Mercury," *Berkeley Daily Gazette*
Science Service (Feb. 4, 1926) "Elements Transmuted," *Berkeley Daily Gazette*
Scientific American (December, 1924) "Why We Are Trying to Make Gold," **131**(6), p. 389
Scientific American (March, 1928) "The Retreat of the Modern Alchemists," **138**(3), p. 208-9
Sclove, Richard E. (Spring, 1989) "From Alchemy to Atomic War: Frederick Soddy's 'Technology Assessment' of Atomic Energy 1900-1915," *Science, Technology, & Human Values*, **14**(2), p. 163-94, Sage Publications, Inc.
Sharma, B.K. (2001) *Nuclear and Radiation Chemistry*, Krishna Prakashan Media
Sheldon, Horton and Estey, Roger S. (Dec. 10, 1926) "Report on the Attempted Change of Mercury Into Gold," *Industrial and Engineering Chemistry*, **4**(23), p. 5-7
Sherr, R., Bainbridge, K.T., and Anderson, H.H. (1941) "Transmutation of Mercury by Fast Neutrons," *Physical Review*, **60**, p. 473-9
Siemens and Halske Aktiengesellschaft (May 7, 1925) "Improvements in or Relating to the Extraction of Precious Metals," U.K. Patent Specification 233,715
Siemens and Halske Aktiengesellschaft (June 12, 1925) "A Process for Converting Mercury Into Another Element," U.K. Patent Specification 243,670
Skinner, Clarence A. (July 1905) "The Evolution of Hydrogen From the Cathode and Its Absorption by the Anode in Gases," *Physical Review*, **21**(1), p. 1-15
Smith, Sinclair (Jan. 1, 1924) "Note on Electrolytically Exploded Wires in High Vacuum," *Proc. of the National Academy of Science: Physics*, **10**(1), p. 4-5
Smits, Arthur (Oct. 25, 1925) "Transformations of Elements," *Nature*, **114**(2869), p. 609-10

Smits, Arthur (Jan. 2, 1926) "The Transmutation of Elements," *Nature*, **117**(2931), p. 13-15

Smits, Arthur (May 1, 1926) "The Transmutation of Elements," *Nature*, **117**(2948), p. 620

Smits, Arthur (Oct. 1, 1927) "Transmutation of Elements," *Nature*, **120**(3022), p. 475-6

Smits, Arthur and Karssen, Albert (Aug. 7, 1925) "Vorlaufige Mitteilung uber einen Zerfall des Bleiatoms," *Naturwissenschaften*, **13**(32), p. 699

Soddy, Frederick (May 5, 1904) *Radio-Activity: An Elementary Treatise From the Standpoint of the Disintegration Theory*, Van Nostrand Co.

Soddy, Frederick (1909) *The Interpretation of Radium*, Putman and Sons, London

Soddy, Frederick (1913) "Radioactivity," *Annual Reports on the Progress of Chemistry*, 10, p. 262-88

Soddy, Frederick (Dec. 4, 1913) "Intra-Atomic Charge," *Nature*, **92**, p, 399-400

Soddy, Frederick and Mackenzie, Thomas (1908) "The Electric Discharge in Monatomic Gases," *Proceedings of the Royal Society of London*, **80**-A, p. 92-109

Srivastava, Yogendra. N., Widom, Allan and Larsen, Lewis (Oct. 2010) "A Primer for Electro-Weak Induced Low Energy Nuclear Reactions," *Pramana - Journal of Physics*, **75**(4) 617-637

Stewart, Alfred W. (1914) *Chemistry and Its Borderland*, Longmans, Green and Co.

Stewart, Alfred W. (1920) *Recent Advances in Physical and Inorganic Chemistry*, 4th ed., Longmans, Green & Co.

Stewart, Alfred W. (1926) *Recent Advances in Physical and Inorganic Chemistry*, 5th ed., Longmans, Green & Co.

Sternglass, Ernest (1997) *Before the Big Bang — The Origin of the Universe*, New York: Four Walls Eight Windows

Strutt, John William, and Ramsay, William (1895) "Argon, a New Constituent of the Atmosphere," *Philosophical Transactions of the Royal Society of London*, **186**, p. 187-241

Stuewer, Roger H. (1986) "Rutherford's Satellite Model of the Nucleus," *Historical Studies in the Physical Sciences*, **16**, p. 321-52

Sunday Times (March 16, 1913) "Coming Doom of Gold"

Taylor, Harold John (1935) "The Disintegration of Boron by Neutrons," *Proceedings of the Physical Society*, **47**(5) p. 873-876

Thomasville Times-Enterprise (Sept. 11, 1958) "World-Famous Scientist to Address Executives," p. 3

Thomson, Joseph John (1897) "Cathode Rays," *Philosophical Magazine*, Series 5, 44, p. 293-316; *Nature* 55, p. 453

Thomson, Joseph John (March, 1904) "On the Structure of the Atom: An Investigation of the Stability and Periods of Oscillation of a Number of Corpuscles Arranged at Equal Intervals Around the Circumference of a Circle; With Application of the Results to the Theory of Atomic Structure," *Philosophical Magazine*, Series 7, **6**, p. 237-65

Thomson, Joseph John (Feb. 13, 1913) "On the Appearance of Helium and Neon in Vacuum Tubes," *Nature* **90** (2259), p. 645-7

Thomson, Joseph John (March 7, 1913) "On the Appearance of Helium and Neon in

Vacuum Tubes," *Science*, **37**(949), p. 360-4

Tilden, William A. (1918) *Sir William Ramsay: Memorials of His Life and Work*, McMillan

Time (Aug. 20, 1945) "Technology Origins"

Times Literary Supplement (June 26, 1903) "The Disintegration Theory of Radioactivity," **76**, p. 201

Times of London (Aug. 25, 1920) "British Association: Cardiff Meeting Opened"

Times of London (Sept. 12, 1933) "Breaking Down the Atom: Transformation of the Elements"

Travers, Morris William (1928) *The Discovery of the Rare Gases*, Edward Arnold

Travers, Morris William (1956) *A Life of Sir William Ramsay*, Edward Arnold

Trenn, Thaddeus (March 1974) "The Justification of Transmutation: Speculations of Ramsay and Experiments of Rutherford," *Ambix*, **21**(1), p. 53-77

Trenn, Thaddeus J., ed. (1975) *Radioactivity and Atomic Theory: Presenting Facsimile Reproduction of the Annual Progress Reports on Radioactivity 1904-1922 to the Chemical Society by Frederick Soddy*, Taylor & Francis, London

Trenn, Thaddeus J. (1981) *Transmutation: Natural and Artificial*, Heyden

United States Geological Survey (Jan. 25, 1922) as quoted by Jewelers Circular

Urutskoev, Leonid, Liksonov, V.I., Tsinoev, V.G. (2002) "Observation of Transformation of Chemical Elements," *Annales Fondation Louis de Broglie*, **27**(4), p. 701-726

Villard, Paul (April 9, 1900) "Sur la Réflexion et la Réfraction des Rayons Cathodiques et des Rayons Déviables du Radium," *Paris Académie des Sciences*

Wendt, Gerald L. (May 10, 1922) "Decomposing the Atom," *Nation*, p. 563-4

Wendt, Gerald L. (May 26, 1922) "The Decomposition of Tungsten," *Science*, **55**(1430), p. 567-8

Wendt, Gerald L. (1939) *Science for the World of Tomorrow*, W. Norton & Co.

Wendt, Gerald L. and Irion, Clarence E. (Sept. 1922) "Experimental Attempts to Decompose Tungsten at High Temperatures," *Journal of the American Chemical Society*, **44**(9), p. 1,887-94

Westwick, Peter J. (2005) "Cherenkov Radiation," in *Oxford Guide to the History of Physics and Astronomy*, Heilbron, J.L., ed., Oxford University Press, ISBN-10: 0195171985, p. 1

Wick, Gian-Carlo (March 3, 1934) "Sugli Elementi Radioattivi di F. Joliot e I. Curie," *Rendiconti Accademia*, Lincei, Italy, **19**, p. 319-324

Widom, Allan and Larsen, Lewis (March 9, 2006) "Ultra-Low-Momentum Neutron-Catalyzed Nuclear Reactions on Metallic Hydride Surfaces," *European Physical Journal C - Particles and Fields*, **46**(1), p.107-10

Winchester, George (1914) "On the Continued Appearance of Gases in Vacuum Tubes," *Physical Review*, **3**(4), p. 287-94

Youth's Companion (Jan. 15, 1920) "Learning About Nitrogen," reprinted in *Plainfield Messenger*

Index

Page numbers in **bold** indicate figures, tables, and photos.

accelerator. *See* particle accelerator
actinium, found in nature, 2
air, 129, 135, 205, 284
   air gap, 205, 207
   air leakage/contamination in experiments, 86, 87, 93, 116, 122, 123–124, 134, 188, 301, 303, 319
   concentrations of gases in, **116**, 135
   liquefied air, 52, 128
Alchemical Society, 146
alchemy, xiv, 5–20, 330. See also *New York Times*, on alchemy and transmutations
   death of modern alchemy (1925-1927), 251–272
   early transmutation research rediscovered, 71–76
   early transmutations dismissed as, 1–4
   paintings and drawings related to, *xviii*, **13**, **73**
   Rutherford fearing his work seen as, 37, 42, 57–58, 60, 68, 148
   Soddy's opinions about, 38–39
   table of alchemical tradition and history, **8**
Allison, Samuel King, 200, 202–206, **203**, **205**, 208, 209, 210, 326
alpha (particle, emission), 60, 138, 147, 148, 158, 222, 237, 321
   discovery of, 28, 35, 331
   as a helium nucleus, 60, 134, 321, 333
   high-energy alpha particles, 137, 272, 290
   man-made transmutation without, 111, 134
   use of by Blackett (1925), 274–293, **275**, **288**, 295–296, 319
   use of by Paneth and Peters, 300
   use of by Ramsay, 79, 112, 149, 315–316
   first use of for man-made transmutation, 3, 78–80
   use of by Rutherford, 185
   comments on Blackett's work in 1925, 294–295
   follow-up to 1919 experiments, 196
   to knock proton off nitrogen (1919), 89, 147, 150–154, 176, 177, 270, 273–274, **275**, 277, 278–280, 283–285, 290, 293, 317
   use of by Wendt, 179
alpha decay, 61, 63
Ambix (publication), 88
American Association for the Advancement of Science, 216
American Chemical Society, xvii, 26, 176, 182, 188, 196, 197, 253, 254, 260–261, 317
American Institute of the City of New York, 181
American Museum of Natural History, 248
American Nuclear Society, xvii
American Physical Society, 263, 265
Anderson, John August, 168, 185, 186, 201
Anderson Benjamin M., 226
anode, 138, 268
antimony, making gold from, 27
Ao, Bing-yun, 323
argon, xv, 40, **54**, 95, 330, 331, 332
   discovery of, 28, 40
   Merton transmutation (synthesis), 133–134

360 • INDEX

Ramsay finding in radium
  dissociating experiments, 85
study of conductivity of (Soddy
  and MacKenzie), 95–96
thorium-to-argon
  transmutation (Rutherford
  and Soddy), 39–42, **41**, 50, 51,
  58
Armstrong, Henry E., 67, 68
artificial gold. *See* gold
asbestos. *See* palladium, effects of
  palladium types in Paneth and
  Peters experiments
Associated Press, 108, **177**, 188, 222,
  223, 253, 258, 306, 307–308
Atkinson, Robert, 335
atom, **32**
  first man-made disintegration
    of a stable atom, 147–155
  indivisibility of (Dalton), 21, 50,
    51, 256, 330, 331, 332
  makeup of an atom, basic
    diagram of, 29–34
  models of, 238, 244, 334
atomic energy. *See* nuclear energy
Atomic Energy Commission, 147,
  149, 216, **276**
atomic mass (AMU), 235, 316
atomic number, 287, 295
atomic science. *See* nuclear physics
  (nuclear science)
atomic structure, models and
  theories. *See* atom, models of
atomic transformation. *See*
  transmutation
"Atoms for Peace" conference, 181

Babson, Roger W., 162–163
Bacon, Francis, 2
Badash, Lawrence, 43–44, 52, 57
Baily, Charles, 238
Bakersfield Morning Echo
  (newspaper), 222, **223**
Baltimore Sun (newspaper), 89, **90**
Baly, E. C. C., 145, 326
Beaudette, Charles, 82
Becquerel, Alexandre Edmond, 25

Becquerel, Antoine Henri ("Henri"),
  24, 25, 196, 270, 330, 331
  and the discovery of
    radioactivity, 23–25, 26, 27–
    28, 128, 330
Bémont, Gustave, 47, 98, 331
Berichte der Deutschen Chemischen
  Gesellschaft. *See* Journal of the
  German Chemical Society
Berkeley Daily Gazette
  (newspaper), 256
beta (particle, emission), 41, 60–61,
  113, 117, 222, 321
  discovery of, 28, 35, 60, 331
  as electrons, 60, 113, 117, 321
beta decay, 61, 63, 338
Bethe, Hans, 336
Bieler, Etienne S., 334
binding energy, liberation of
  excess, 335
Birla Temple (New Delhi), **14**, 14–
  18, **15**, **16**, **17**
Blackett, Patrick Maynard Stewart,
  274–275, 296, 335
  discovery of first artificial
    transmutation of a stable
    atom, 290
  nitrogen-to-oxygen
    transmutation, xvi, 147, 153–
    154, 273–296, **275**, **288**, 297,
    319, 335
    Rutherford's critique of,
      291–293
    seeing process as
      integration, 281, 292–
      293
Bockris, John O'Mara, 323
Bohr, Niels, 150, 238, 269–270, 333,
  334
Bolton, H. Carrington, 20, 26–27
Boltwood, Bertram Borden, 93
boron, 152, 153, 292
  boron-10, 336
Boston Globe, 159
Bowman, Robert C., Jr., 323
Boyle, Robert, 19
Bragg, William, 247

Brahe, Tycho, 9–10
Brice, Edward C., 27
Brillouin, Louis "Marcel," 247
Briscoe, Henry Vincent Aird, 200, 206–209, **207**
British Association for the Advancement of Science, 66, 92, 95, 105, 108, 109, 157, 160, 161, 334
British Chemical Society, 206
Burbidge, Geoffrey and Margaret, 336

Cambridge University, 35, 113, 125, 130, 175, 279–280
Cameron, Alexander Thomas, 2, 82–83, 85–87, **86**, 89, 91, 93, 103–104
Campbell, John, 279
  Rutherford: Scientist Supreme (Campbell), 277
Canton Times (newspaper), 159
carbon, 92–93, 121, 153, 232, 241, 243, 292
  carbon-12, 31, **32**
  carbon-13, 31, 152
carbon dioxide, 2–3, 188
carbon disulphide, 271
carbon monoxide, 122
Cardoso, E., 3, 145, 326
Castaigne, André
  "M. Pierre and Mme. Marie Curie Finishing the Preparation of Some Radium" (illustration by André Castaigne), *iv*
cathode rays, 3, 117, 128, 129, 132, 143, 176–177
  cathode-ray tube, 111–112, 117–118, **120**, 145–146, 175, 183, 316
    confirming elemental transmutations, 127, 182–183
    denying elemental transmutations, 132–133
    rare gases research, 119–138, 140, 142
  and discovery of electrons, 125, 331
Cavendish laboratory, 35, 113, 201–202, 272, 274, 279–280, 290, 335
CBS-TV, 180
Chace, William G., 167-168, 173, 201
Chadwick, James, 284–285, 286, 327
  confirming existence of neutrons, 244, 250, 313, 319, 334, 335
charged particle, 337
Charleston Daily (newspaper), 223
Chase National Bank of New York, 226
Chemical Institute (University of Berlin), 258, 260, 263, 266, 297, 311
Chemical News (publication), 43–44
Chemical Society, 43, 89, 91, 95, 118, 120–121, **122**, **123**, 139, 206
Chien, Chun-Ching, 323
Clarendon Laboratory, 137, 267–268, 318
Clarke, Arthur C.
  Profiles of the Future (Clarke), 104
Close, Frank, 312
  Too Hot to Handle: The Race for Cold Fusion (Close), 312
coal reserves in U.K., Ramsay foreseeing depletion of, 105–108
  500-year shift of energy sources in the U.K., **108**
  graph of annual U.K. coal production and imports, **106**
Cockcroft, John, 335
"cold fusion," xvi, 312, 330
collective effects, 337
Collie, John N., 3, 38, 118–119, **119**, 146, 316, 333
  on occlusion hypothesis, 176, 194, 333
  rare gases research and transmutation, 113, 118–137, 139, 145, 183, 289, 316, 325, 326, 333
Collins, Harry M.
  Changing Order: Replication and

362 • INDEX

Induction in Scientific Practice (Collins), 81
Congress of the International Union of Pure and Applied Physics, 247
copper, 10, 84, 85, 146
   copper-to-lithium transmutation, 2, 85, 89, 91, **92**, 92, **92**, 139
copper nitrate, 78, 85
copper oxide, 136
copper sulfate, 78, 84–85, 89
Cornell Baker Laboratory, 307
Cornell Daily Sun (newspaper), **307**
Cornell University, 9, 11, 307, 308, 311, 313
cosmochemistry, 297
Costa, Mariagrazia
   The Lost Elements: The Periodic Table's Shadow Side (Fontani, Costa, and Orna), 74–75
Coulomb barrier, 244, 335
Crookes, William, 43–44
Crookes tube, 21, 95, **96**
Curie, Marie Sklodowska, **iv**, **48**, 270
   attempt to replicate Ramsay's experiment, 2, 83, 85–86, 87, 91, 129–130, 277
   and radioactivity, 26, 27–28, 179, 196, 331
   and radium, **iv**, 46, 47–50, 98, 142, 331
Curie, Pierre, 27–28, **48**, 270
   and radium, **iv**, 46, 47–50, **49**, 51, 98, 142, 331, 332

Dahl, Per Fridtjof
   From Nuclear Transmutation to Nuclear Fission, 1932-1939 (Dahl), 294–295
Daintith, John
   Biographical Encyclopedia of Scientists (Daintith), 282
Dalton, John, 19–20, 270
   New System of Chemical Philosophy (Dalton), 20
   theory of the indivisible atom, 21, 50, 51, 256, 330, 331, 332
Darwin, C. G., 334
Dash, Vaidya Bhagwan, 15
   Alchemy and Metallic Medicines in Ayurveda (Dash), 14
decay. *See* radioactivity, radioactive decay
deuterium, 29–30, 332
deuteron, 335
Deutschen Chemischen Gesellschaft. *See* Journal of the German Chemical Society
disintegration, atomic, 99, 110, 183, 201, 204, 246, 290, 335. *See also* radioactivity, radioactive decay
   and Blackett's work with nitrogen-to-oxygen, 285–296, 297, 335
      seeing as integration rather than disintegration, 281, 282, 292–293
   Disintegration Theory (Soddy), 42, 44, 57, 59–60, 65–69, 77, 332
   energies involved in, 3, 4, 103–104, 161–162, 194, 315
   first observed nuclear disintegration, 35–46, 273–274, 331
   man-made disintegration of a stable atom (Rutherford), 147–155, 157, 160, 165, 189, 196, 256, 270, 290, 307, 317, 327, 334
   natural disintegration, 38, 131, 203, 204, 209, 232, 246–247, 256, 290
      first observed, 35–46, 273–274, 331
      Rutherford's concept of, 35–46, 57, 68, 69, 77, 89, 331
      Soddy's concept of, 35–46, 103–104, 290, 331
   and radium, 48, 50
   and Ramsay, 77–93, 117

Rutherford nitrogen-to-oxygen
myth, 273–274, 279, 280–282,
283–284, 287, 289, 291, 297
versus transmutation, 38, 43–44
and Wendt, 194, 196, 203, 209
Duchesne, Joseph, 71
Duhme (researcher), 265
duplicating experiments. *See*
scientific method and
repeatability

Egan, Mortimer A., 199–200
Egerton, Alfred Charles Glyn, 3, 137–138, 148, 326
Einstein, Albert, 50, 78, 98, 109, 203, 213, 332, 336
Eisenhower, Dwight D., 215–216
electrically dominated LENR reactions, 337–338
The Electrician (publication), 66–67, 68
electrolysis, 268
electromagnetic force, 33
electromagnetic spectrum of visible light, **53**
electrons, 22, 29, **32**, 147, 154, 169, 185, 195, 216, 332, 334
   addition or subtraction of causing changes, 33, 152, 287, 295
   as beta particles, 60, 113, 117, 321
   and cathode-ray tubes, 117, 125, 128, 132, 175
   discovery of, 26, 125, 130, 331
   and efforts to make gold, 232, 235, 252, 253, 265, 268
   electron capture, 334, 336
   and LENRs, 338
   low-energy electrons, 175, 272
electroweak interactions, xvii, 4, 337. *See also* low-energy nuclear reactions (LENRs)
elements, 29–34. *See also* isotope
emanations
   and radioactive elements, 28, 42, 58, 66–67
   radium emanations, 2, 46, 52, 58, 64, 84, 91, 93, 283, 332
   radon emanations, 85
   Rutherford on, 2–3, 65, 66–67
   thorium emanations, 37, 40, 41, 42, 43, 58, 64, 68, 331
emissions. *See* alpha (particle, emission); beta (particle, emission); gamma rays (gamma radiation, gamma emission)
Emmens, Stephen H., 27
energy
   energy sources in the U.K., 105–108, **106**, **108**
   Foote on transmutation of elements creating, 214, **215**
   KeV energy, 336
   mass deficit when transmuting up the periodic table, 216–217
   Rutherford on no future for nuclear energy, 108–110, **109**
   Soddy seeing radioactivity as energy source, 97–100, 105
English Mechanic and World of Science (newspaper), 157–158, **158**
Estey, Roger Shephard, 230, **230**, 233, **233**, 233, 258–263, 265, 269
Europe and alchemy, 8–11
Eve, Arthur, 68, 93, 195–196
Exeter College. *See* Clarendon Laboratory
exploding electrical conductors (exploding-wire phenomenon), xv–xvi, 167–174, **172**, **173**, **174**, **184**, **205**
   University of Chicago experiment of Wendt and Irion, 175–198, **205**, 317, 334
   criticisms and efforts to repeat, 192–198, 199–211, 249, 295, 317
   Urutskoev, confirmation by, xvi, 208
F. A. Paneth Meteorite Trust, 299
Fernandez, Bernard, 296
"First Law" of Arthur Clarke, 104
Fisher, Irving, 162, 163, 164–165

364 • INDEX

fission. *See* nuclear fission
Fleischmann, Martin, xvi, 89
fluorine, 275, 287, 289, 294, 335
Fontani, Marco
   The Lost Elements: The Periodic Table's Shadow Side (Fontani, Costa, and Orna), 74–75
Foote, Paul Darwin, 213–216, **216**, 262
force. *See* electromagnetic force; strong force; weak force and weak interactions
Foucault, Michael, 25
Fouquet, Roger, 107
Fowler, William, 336
Franklin, Benjamin, 169–170
Free, Edward E., 222, **222**, 222
Freeman, John
   Suppressed and Incredible Inventions (Freeman), 72
French Academy of Science, 24, 25
Freund, Ida, 51
Frisch, Otto, 98, 336
Früngel, Frank, 170–172
Fry, Al, 72
fusion. *See* nuclear fusion
Futagami, Tetsugoro, 201

Gaedicke, Johannes, 217
Galison, Peter, 279–282
gamma rays (gamma radiation, gamma emission), 60–61, 241, 321
   identifying of, 28, 331
   and LENRs, 337, 338
Garrett, Milan Wayne, 267–268, 318
gas behavior in metals analysis, references for, 323
Gaschler, Alois, 268–269
gases, concentrations of in air, **116**, 135
Geochimica et Cosmochimica Acta (journal), 299
German Chemical Society, 254, 299
Gibbs, Philip, 24–25
Giunta, Carmen, 329
Gleditsch, Ellen, 2, 83, 85
gold, xv, 146, 147
   antimony-to-gold transmutation, 27
   gold-to-mercury transmutation, 262–263, 268–269
   impact of making artificial gold, 163, 222, **224**, 224–226, **225**, 226, 230–231, 252
   mercury-to-gold transmutations, 317–318
      efforts to recreate Miethe-Stammreich research, 222–225, **223**, **224**, 226–235, **227**, **230**, **233**, 237, 250, 251–279, **259**, 297, 318
      by Miethe-Stammreich, 217–225, 237, 251, 252–256, 257, 297, 306, 317–318, 334
      by Nagaoka, 237–250, 251, 253, 255, 256, 266, 272, 314, 318, 335
   question of presence of gold in mercury, 260, 261, 265, 266
   rumors that Rutherford turned lead into, 157, 159, 162–165
Goldschmidt, Frieda, 258, 260
Gozzi, D., 323
Gulf Oil Corporation, 214–215

Haase, Wilhelm, 258, 260, 266
Habashi, Fathi, 24
Haber, Fritz, 221, 232–233, 253–255, **264**, 264–268, 314, 318
Hahn, Otto, 98, 336
half-life, 62, 127, 235, 331
Hansen, Steve P., 168–171
   belljar.net Web site, 167
Harkins, William Draper, 200, 202–206, **203**, **205**, 209, 232, 295–296, 326
Harper's Weekly (magazine), 19
Harrow, Benjamin, 88
Harteck, Paul, 127
Hasselberg, K. B., 59–60
heavy hydrogen. *See* deuterium
helium, **54**, 95, 116, 203, 330, 333

INDEX • 365

alpha particle as helium
  nucleus, 60, 134, 321, 333
  and "alpha" rays, 331, 332, 333
  as a byproduct of
    disintegration, 67
  comparison with hydrogen in
    hydride-forming metals, **114**
  discovery of, 28
  helium-3, **30**, 30–31, **31**, 126–129,
    335
  helium-4, xv, 30–31, **31**, 127, 336
  helium transmutation
    (synthesis), 38, 111, 137-138,
    333
      by Collie and Patterson,
        113, 118-125, 134-137
      by Masson, 122
      by Paneth and Peters, xvi,
        299, **299–314**, **307**, 319
      by Ramsay, 82-85, 117–118
      by Ramsay and Collie, 113
      by Ramsay and Soddy,
        51–56, 59, 332
      by Soddy and MacKenzie,
        95-96, 112-113
      by Thomson, 113, 125-129
      by Wendt and Irion, xv,
      by Winchester, 131-133
  helium transmutation
    (synthesis) denial by Wendt,
    183
  helium transmutation
    (synthesis) failures
      by Curie, 129
      by Thomson, 130
      by Egerton, 137
  myth that Rutherford
    transmuted hydrogen into
    helium, 154–155, 175
  and occlusion, 96, 332, 333
  permeation (or solubility) in
    metals and glass, 114–116, **115**,
    145
  presence of in exploding-wires
    experiment (Wendt and Irion),
    187–188
  Ramsay finding in radium
    dissociating experiments, 85
  references for gas behavior in
    metals analysis, 323
  study of conductivity of (Soddy
    and MacKenzie), 95–96, 112–
    113
  tungsten-to-helium
    transmutation by Wendt and
    Irion, 175–198, 199
      Allison and Harkin's
        failure to replicate,
        202–206
      Briscoe's failure to
        replicate, 206–210
      replication failures, 199–
        211
      Smith's failure to
        replicate, 201–202
      X-ray tubes not producing
        helium normally, 194–195
Henry VI (king of England), 10
Herdman, W. A., 160
high-energy physics, 33, 80, 88, 111,
  240, 244, 315. See also alpha
  (particle, emission)
  high-energy methods for
    transmutations, 134, 137, 182–
    183, 272, 290, 297, 313, 315,
    316, 337–338
high-voltage discharge
  experiments, 111, 112, 168, 316
Hirshberg, Leonard Keene, 19, 129–
  131
Hodko, Dalibor, 323
Hofmann, K. A., 221, 260
Hönigschmid, Otto, 252–253
Houtermans, Fritz, 335
Howorth, Muriel, 40, 41, 45–46, 67–
  68
Hoyle, Fred, 336
Hu, Wang-yu, 323
The Humanist (publication), 181
Hunter, Rudolph Melville, 146
hydride-forming metals, **114**, 114–
  116
hydrogen, **54**, 335. See also
  deuterium; proton; tritium

366 • INDEX

claimed impossibility of helium disintegrating into hydrogen, 203
comparison with helium in hydride-forming metals, 114
as a component of nitrogen according to Rutherford, 150–152
hydrogen-hydrogen fusion, 336
myth that Rutherford transmuted hydrogen into helium, 154–155, 175
myth that Rutherford transmuted nitrogen to hydrogen, 157–162
occluded in metals, 332
proton (claimed as hydrogen) emission from nitrogen disintegration by Rutherford, 150–152, 157–159
protons first seen by Rutherford as a hydrogen particle, 155, 158, 177, 284, 294, 317, 334
    proving many elements composed of "hydrogen," 154–155, 177–178
    scientists thinking all elements made of hydrogen particles, 217
radium dissociating hydrogen and oxygen, 84–85

Illustrated World, 199–200
The Independent (newspaper), 277, 278
India and alchemy, 14–18
Institute of High Energy Physics (IHEP)
    "Chronology of Milestone Events in Particle Physics" (IHEP), 329
Institute of Physical and Chemical Research, 201, 238, 258, 318
integration, atomic, 183
    Blackett recognizing, 281, 282, 292–293

ion, 285
ionization, 185
Iowa City Press-Citizen (newspaper), 219–220
Irion, Clarence E., 80, 244, 249
    University of Chicago exploding-wire experiment, 175–198, **205**, 317, 334
        efforts to replicate, 198, 199–211, 295, 317
iron, 19, 104, 137, 146, 232
    elements heavier than, 336
    exploding iron alloy wire, 171, **172**
isotope, 30–33, 62
    Blackett's transforming nitrogen to an oxygen isotope, 274, 281, 282, 287, 293, 295
    and LENRs, 338
    Rutherford's hypotheses about, 152, 154
    Soddy developing concept of, 69, 103, 333
    study of mercury isotopes leading to gold, 237–250
    unstable isotopes, 127, 235, 244, 338, 341
isotopic abundance, 31

Jaenicke, A., 220–221
"Japanese Einstein." See Nagaoka, Hantaro
Joliot-Curie, Frédéric and Irène, 336
Journal of Alchemical Society, 146
Journal of the American Chemical Society, 72, 193, 196, 200
Journal of the Chemical Society Transactions, 200, 292
Journal of the German Chemical Society, 299, 306, 313
Journal of the Mineralogical Society of America, 248, 249
Journal of the Radiological Society of North America, 181

Kahlbaum company and

palladium-asbestos, 306, 308, 309
Kansas City Star (newspaper), 258, 268–269
Karssen, Albert, 256–258, **257**, 263, 265, 271, 318, 335
Kelvin, Lord (William Thomson), 67, 68
Kendall, James, 161, 162
Kent Chemical Laboratory, 176, 200, 202, 232, 234–235, 255, 258, 295, 318
Knights Templar, 8
Krivit, Steven
   New Energy Times Web site, 76, 321
krypton, 28, 331
Kunz, George Frederick, 246, **248**, 248–249
kunzite (gem), 249
Kurlbaum (German physicist), 252

Laborde, Albert, 49, 51, 332
Lakshmi Narayan Temple. *See* Birla Temple (New Delhi)
The Lancet (journal)
   "Modern Alchemy: Transmutation Realized" (The Lancet), 91
Larsen, Lewis, 4, 243–244, 246, 249–250, 320
   Widom-Larsen theory, 320, 338
Law of Radioactive Change (Rutherford and Soddy), 56, 58–59, 60, 61–62, **62**, 64–65, 66, 69, 332
Lawson, Robert W., 145, 326
lead, 40, 100, 143, 189, 192, **257**
   lead-to-gold transmutation, 3, 7, 182, 213, 216, 217, 220, 263
      rumors of Rutherford accomplishing, 157–165
   lead-to-mercury transmutation, 256, 257, **257**, 263, 268, 271, 318, 335
   lead-to-thallium transmutation, 263, 268, 318, 335
   and X-rays, 23, 128
Le Bon, Gustave, 25
Leech, Paul N., 188–189
Le Matin (publication), 196, **197**, 197–198, **278**, 278–279
LENRs. *See* low-energy nuclear reactions (LENRs)
light, full spectrum of visible light, **53**
Lindemann, Frederick Alexander, 267
Lippes, Lorna, 329
lithium, 78, 85, 332
   lithium-7, 336
   Ramsay's copper-to-lithium transmutation, 89, 91, **92**, 92
      Curie's attempt to replicate, 2, 83, 85–86, 87, 91, 129–130, 277
Lodge, Oliver Joseph, 67, 142–143, 291
Loleit, H., 313
London Institution, 55
Lorentz, Hendrik Antoon, 247
Lo Surdo, A., 146
Lotz (researcher), 265
low-energy electrical discharge experiments, 336
low-energy methods for transmutations, 113, 133, 138, 175, 271–272, 297, 314, 315–320
low-energy nuclear reactions (LENRs), xvii, 133, 337–338

MacKenzie, Thomas D., 95–96, 112–113, 332, 333
Madison Daily Herald, **177**
magnetically dominated LENR reactions, 338
Manhattan Project, 202, 336
man-made disintegration of a stable atom. *See* disintegration; disintegration, atomic
man-made transmutation, 38, 81–82, 332, 333, 334–335. *See also* media coverage of transmutation experiments; nuclear

368 • INDEX

transmutation
  by artificially accelerated particles, 335
  distinctions between man-made and natural, 111
  man-made nuclear transmutation, 38, 59, 139, 147, 273, 276, 289, 296, 316, 333, 334, 335
    experiments giving evidence of rare gases, 111–138
  and Ramsay's work, 77–93, **86**, **90**, **92**, 316
  Rutherford's 1919 work labeled as, 160, 189
  scientists unwilling to support idea of, 78, 80
Marsden, Ernest, 283–284
Marum, Martin van, 168
Mason City Globe Gazette (newspaper), 189–190
mass-3, 152. *See also* tritium
Massachusetts Institute of Technology (MIT), 173, **173**, 262
Masson, James Irvine Orme, 3, 316, 325, 326
  experiments giving evidence of rare gases, 122–123, 124, 135–136, 137, 145, 333
  testing occlusion hypothesis, 176, 194, 333
Max Planck Institute for Chemistry, 298
McGrath, James R., 167-168
McLean, Adam, 6
  Alchemy Web Site, 7, 80
media coverage of transmutation experiments, 88–93, **90**, **92**, 139–146, **149**, 157–165, 159, 278–279. *See also* names of specific newspapers, magazines, etc.,
Meitner, Lise, 98, 336
Mellon Institute of Industrial Research, 215
Mellor, Joseph William, v, 1–4, 146
  A Comprehensive Treatise on Inorganic and Theoretical Chemistry (Mellor), 1
Merchel, Silke, 298
mercury, 10, **54**, 128–129, 204, 237, 271, 335
  gold-to-mercury transmutation, 262–263, 268–269
  lead-to-mercury and thallium transmutation, **257**, 263, 268, 318
  mercury-to-gold transmutations, 217, 251–252
    Birla Temple (New Delhi), 18
    efforts to recreate Miethe-Stammreich research, 222–225, **223**, **224**, 226–235, **227**, **230**, **233**, 237, 250, 251–279, **259**, 297, 318, 335
    by Miethe-Stammreich, 217–229, 237, 251, 252–256, 257, 297, 306, 317–318, 334
    by Nagaoka, 237–250, 251, 253, 255, 256, 266, 272, 314, 318, 335
  presence in Wendt-Irion experiments, 187
  question of presence of gold in, 260, 261, 265, 266
Merton, Thomas Ralph, 3, 133–134, 137, 145, 325
metals analysis, references for gas behavior in, 323
metals analysis, references for helium in. *See also* hydride-forming metals
Miethe, Adolf, **219**, 266, 269, 270, 272, 314, 334
  mercury-to-gold transmutations, 217–229, 237, 252–256, 257, 297, 306, 317–318, 334
    efforts to recreate or discredit research, 222–225, **223**, **224**, 226–235,

227, 230, 231–233, **233**, 237, 250, 251–279, **259**, 297, 318
    Siemens and Halske acquiring rights to research, 251–252
Miles, Melvin, 115–116
Millikan, Robert Andrews, 103–104, 247, 334
    Science and the New Civilization (Millikan), 103–104
Minevski, Zoran, 323
Mining and Scientific Press (journal), 162
Mladjenovic, Milorad, 88, 283–284, 285, 289, 293
    The History of Early Nuclear Physics (Mladjenovic), 88, 283–284
Montreal Gazette (newspaper), 238, 246–247, **247**, 249
Morning Post (newspaper), 121–122, 139, 141–142, 237, 306
Morrisson, Mark, 80
    Modern Alchemy: Occultism and the Emergence of Atomic Theory (Morrisson), 80
Mount Wilson Observatory, 168, 185–186, 200, 201

Nagaoka, Hantaro, 80, 201, 235, **239, 240, 241**, 258, 263
    attempt to develop a model of the atom, 238, 244
    mercury-to-gold transmutations, 237–250, 251, 253, 255, 256, 266, 272, 314, 318, 335
Nairne, Edward, 170
The Nation (magazine), 183, 185, 187, 192, 196, 197
National Academy of Sciences, 209, 295–296
National Agency for International Publications Inc., 181
natural nuclear disintegration. *See* disintegration, atomic

natural nuclear transmutation, 38, 42, 59, 77, 100–101, 161–162, 290
    acceptance of by scientists, 77
    distinctions between man-made and natural, 111
    proof of (Ramsay and Soddy), 51–56, 68, 77
Nature (journal), 72, 145
    on mercury-to-gold transmutation efforts, 217, 220–222, 225, 237–238, 241, 244, 249–250, 251, 253, 260, 265–266, 268, 270
    on Paneth and Peter's hydrogen-to-helium transmutation, 306
        Paneth submitting a partial retraction to, 310–311, 313
    "The Present Position of the Transmutation Controversy" (*Nature*), 265–266, 268, 270
    and Ramsay, 92, 113, 117, **117**, 118
    and Rutherford, 154–155, 193, 196, 197, 286
    on Blackett, 291, 292
Naturwissenschaften (journal), 218, 219, 220–222, 237, 250, 252–253, 263, 266, 267, 299, 300–301, 302, 304–306
Nelson, Robert, 71–76, 80, 320, 329
    Adept Alchemy (Nelson), 72, 75, 220
    commissioning a drawing "Alchemia," **73**
neon, xv, 92, 114, 122, 124, 132–133, 146, 175, 201, 316
    and argon, 134
    atmospheric contamination, 116, 301, 303
    Collie and Patterson's work with, 3, 38, 118, **119**, 119–138, **120**, 333
    discovery of, 28, 78, 83, 119, 331
    not finding occluded neon, 333
    permeation (or solubility) in metals and glass, 145

370 • INDEX

Ramsay's work with, 38, 78, 83, 85–87, 91, 93, 117, 119, 332, 333
ratio of nitrogen to neon in air, 135
Soddy and MacKenzie's work with, 95–96, 112–113
Thomson's work with and belief in occlusion, 113, 126, 127, 129, 132, 137–138, 143, 175–176
neutrons, 29–30, **31**, **32**, 334, 336
Chadwick verifying existence of, 244, 250, 313, 319, 334, 335
differing number of, creating isotopes, 30, 32–33
before discovery of, 26, 150, 153, 216, 244
fast neutron, 244
and LENRs, 337–338
neutron-capture, 216, 336, 338
thermal neutron, 336
use of fast neutrons, 244
New Energy Times Web site, 76, 321
Newton, Isaac, 19
New York Academy of Sciences, 20
New York Times (newspaper), 77, 80, 132, 143–144, 214
on alchemy and transmutations, 26–27, 100–101, 139, 146, 147, 160–162, 162, **164**, 164–165
Foote on potential of transmutation, **214**
mercury-to-gold transmutations, **253**, 253–254, 258–260, **259**, 261, 264–265
"The Revival of Alchemy" (*New York Times* article), 26–27
tungsten-to-helium transmutation, 182, **182**, 188–189
on artificial gold, **224**, 224–226, **225**
on Ramsay, 55–56, **56**, 89, **90**, 91, 92, 117, 139, 143–144
on U.K. coal production, 105, 106–107
New York Tribune (newspaper), 162
New York University, efforts to recreate Miethe-Stammreich transmutation of gold, 222–224, 229–234, **230**, **233**, **259**, 261–262, 263, 265, 269
Scientific American magazine sponsoring efforts, 222, 226–234, **227**, 237, 258–260, 261–262, 271, 297, 318
Niépce de Saint-Victor, Abel, 24–25, 128, 320, 330
Nieuwenhoven Mat, 205
nitrogen
and air leakage, 116, 123–124, 135
nitrogen-to-hydrogen, 150–152, 157, 158, 159
nitrogen-to-oxygen transmutation/disintegration
Blackett experiments in 1925, 273–296, **275**, **288**, 335. *See also* Blackett, Patrick Maynard Stewart
Rutherford experiment conducted in 1919, xvi, 147, **148**, 150–151, 152–153, 160, 217–218, 334
ratio of nitrogen to neon in air, 135
Rutherford's publications on nitrogen disintegration, 327
Nobel Prize
in chemistry, 28, 44, 46, 57, 59, 65, 69, 78, 103, 195, 290
in physics, 25, 28, 125, 238, 247, 289, 290, 296
Nordmann, Charles, 196, 197, 278–279
Noyes, William Albert, 45–46
nuclear chain reaction, 319, 335, 337
nuclear chemistry, 6, 103, 202, 298
nuclear disintegration, first observed. *See* disintegration,

INDEX • 371

atomic
nuclear energy, 26, 46, 47, 97, 98, 191, 332
   Rutherford not seeing a future for, 108–110, **109**
nuclear fission, 6, 26, 98, 106, 319, 336, 338
nuclear fusion, 26, 315, 335, 336, 337
   "cold fusion," xvi, 312, 320
nuclear physics (nuclear science), xiii–xvii, 4, 5–6, 103, 270, 289, 319–320
   timeline of key events in early nuclear history, 329–336
nuclear processes. *See* neutrons, neutron-capture; nuclear fission; nuclear fusion; weak force and weak interactions
nuclear radiation. *See* radioactivity (radiation, radioactive emissions)
nuclear reaction, 60
   high-energy stimuli needed for, 98–100, 117, 170
   initiated by chemistry, 114–115, 320
   and isotopic shifts, 32
   low-energy nuclear reactions, 74, 133, 320, 337–338
nuclear science. *See* nuclear physics (nuclear science)
nuclear transformation. *See* nuclear transmutation
nuclear transmutation, xv–xvi, 26, 29–30, **30**, 72, 74–75, 80, 332
   high-energy nuclear transmutations, 297, 313, 315. *See also* high-energy physics
   low-energy nuclear transmutations, 133, 271, 297, 314, 315–316. *See also* low-energy methods for transmutations
   man-made nuclear transmutation, 38, 59, 111–138, 139, 147, 273–274, 276, 289, 294, 296, 316, 319, 334, 335
   minimal coverage of in *The Lost Elements*, 74–75
   natural nuclear transmutations, 38, 59, 77, 290
   Soddy seeing potential of, 95–104
   spontaneous nuclear transmutations, 77
   timeline for, 329–336
nuclei. *See* nucleus
nucleosynthesis
   nucleosynthetic reaction network, 63, 63, **63**, 63
   in stars, 336
nucleus, 29–30, **30**, **32**, 33, 187, 269, 332
   alpha particle as helium nucleus, 60, 134, 321, 333
   changes to, requiring high energies, 3, 187, 204, 232, 235, 237
   identifying of, 5, 38
   and isotopes, 30–31, **31**
   Saturn Model of (Nagaoka), 244
nuclide, 38

occlusion hypothesis, 3, 95–96, 124, 135–136, 143, 199–200, 201, 204, 309–310, 323, 332, 333
   occluded gold hypothesis, 266–268
   of Thomson, 113, 126, 127–129, 133, 143, 175–176, 289
      disproving of, 113, 143, 176, 194, 204, 333
Ojasoo, Tiiu, 25
Olean Evening Herald (newspaper), 162, **163**
Oliphant, Mark, 127, 335
Orna, Mary Virginia, 74–75
   The Lost Elements: The Periodic Table's Shadow Side (Fontani, Costa, and Orna), 74–75
Osaka University, 238
Osborne Naval College, 280
Ostade, Adrian van, painting "The

Alchemists," 13
Oxford University, 35, 38, 103, 133, 137, 267–268, 318
Oxford University Press, 74
oxygen, 29, 104, 116
    Blackett's nitrogen-to-oxygen transmutation, xvi, 147, 153–154, 273–296, **275**, **288**, 297, 319, 335
        Rutherford's critique of, 291–293
        seeing process as integration, 281, 292–293
    Collie-Patterson's work with, 121–122, 124, 136
    myth that Rutherford transmuted nitrogen-to-oxygen, xvi, 147, **148**, 150–153, 160, 176–177, **177**, 273, 287, 289, 292–295, 317, 319, 334, 335
    Paneth-Peters' work with, 301–302, 304, 305, 308–309, 311, 312, 319
    Ramsay's work with, 84–85, 333

"packing effect," 203
palladium, 300, 301–302, 303–304, 319
    effects of palladium types in Paneth and Peters experiments, 304–305, 308–309, 309, 311–312
Paneth, Friedrich Adolf (Fritz), 80, 111–112, 297–299, **298**, 335
    "Ancient and Modern Alchemy" (lecture by Paneth), 9–11, 307
    hydrogen-to-helium transmutation, xvi, 11, **299**, 299–314, **307**, 319, 335
        partially retracting their research results, 308–314, 319
Paracelsus (Philippus Aureolus Theophrastus Bombastus von Hohenheim), 11–12
particle accelerator, 6, 138, 329, 335, 337, 338
particles. *See* alpha (particle, emission); beta (particle, emission); charged particle; gamma rays (gamma radiation, gamma emission)
    subatomic particles. *See* electrons; neutrons; protons
Passell, Thomas O., 323
Patterson, Hubert S., 118–119, 289, 316, 325, 326, 333
    hydrogen-to-helium and neon transmutation, 3, 38, 113, 118, 120–125, 127, 135–136, 137, 139, 145, 333
        efforts by others to replicate, 122–123, 136, 333
        Wendt not accepting results, 183
    on occlusion hypothesis, 176, 194, 333
Perman, E. P., 2
Peters, Kurt Gustav Karl, 11, 80, 111–112, 297
    hydrogen-to-helium transmutation, xvi, **299**, 299–314, **307**, 319, 335
        Paneth partially retracting their research results, 308–314, 319
Philosopher's Stone, 7, 11, 161
Philosophical Magazine, 151, 238, 277
photography
    and the discovery of radiation, 21–28
    Miethe's use of mercury vapor photography lamps, 217–225, 317–318
photon, 60
Physical Review (journal), 72
Piutti, A., 3, 326
Place, Robert, 8, 11–12
    Magic and Alchemy: Mysteries, Legends and Unexplained Phenomena: (Place), 6–7

plasma, 168, 169-174, 337, 338
platinum, xv, 127, 235, **257**, 271, 304
  and gold, 241, 243, 253, 255, 257, 268, 317–318, 334, 335
polonium, 2, 28, 35, 331
Pons, Stanley, xvi, 89
Pontanus, J. J., 1
Popular Science Monthly (magazine), 183–184, **184**, 189, **190**, 190–191
Princeton University, 313
Proceedings of the National Academy of Science, 200
protons, 26, 29–31, **30**, **31**, 32–33, 108, 153, 155, 216, 237, 244
  Blackett's study of ejection of protons, **275**, 281, 282, 287, **288**, 289, 291, 292–293, 335
  first seen by Rutherford as a hydrogen particle, 155, 158, 177, 284, 294, 317, 334
  Rutherford's study of proton emission after bombardment, 89, 150, 152–154, 158–159, 176–177, 213, 217, 273, 276, 278–279, 290, 296, 307, 317, 334
    his hypotheses about, 274–275, **275**, 281, 282, 284, 285, 289, 293–294
  in stars, 336
  use of to bombard gold, 268–269
"protyle" concept, 330, 334
Prout, William, 330, 334
puffery, 12, **13**

quantum mechanics, 60, 337
quicksilver. *See* mercury

radioactivity (radiation, radioactive emissions), 2, **24**, 195–196, 269–270, 331
  artificial radioactivity, 336
  discovery of radiation, 21–28, 35, 128, 320, 330
  discovery of radioactivity, 23, 26
  emanations from radioactive elements, 28, 42, 58
  emissions from spontaneous radioactivity, 321
  Law of Radioactive Change, 56, 58, 59, 60, 61–62, 65, 66, 69, 332
    as a manifestation of subatomic chemical change, 58
  Niépce's work on, 24–25, 330
  radioactive decay, 61, 63, 331. *See also* alpha decay; beta decay; disintegration; half-life
    graph of thorium-X decay, **62**, 62
  research on becoming nuclear physics, 5
  Soddy's interpretation of, 97–98
  types of, 60–61. *See also* alpha (particle, emission); beta (particle, emission); gamma rays (gamma radiation, gamma emission)
radiochemistry, 28, 297–298
radium, 2–3, 34, 101, 111, 140, 142, 179, 200, 247
  Curies' study of, **iv**, 28, 35, 47–50, 51, 98, 129, 331, 332
  Hirshberg questioning natural disintegration of, 130–131
  Marsden's work with, 283–284
  radium emanations, 2, 46, 52, 58, 64, 84, 91, 93, 283, 332
  Ramsay and Cameron work with, 82–83, 84–85, 89, 91, 93
  Ramsay and Soddy's work with, 51–56, 59, 66, 68, 71, 77, 84, 93, 112–113, 130, 332, 333
  Rutherford and Soddy's work with, 46, 50, 57–58, 61, 64, 66–67, 93, 150, 290
    Radium C, 89
  Soddy's interpretation of, 66–67, 97–100
  Wendt and Irion's work with, 188–189, 191–192

374 • INDEX

Radium Institute of Vienna, 113
radon, **63**, 63, **63**, 63, 85, 331. *See also* thorium, thorium-to-thorium-X
Ramsay, William, 2, **79**, 139, 149, 316, 323, 325, 330, 331, 332
   discovery of noble gases, 28, 40, 78, 83, 119, 331
   foreseeing depletion of coal reserves in the U.K., 105–108, **106**
   and man-made transmutations, 77–93, **86**, 95
      claim of first, 112, 189, 277, 289, 315–316
      historical and media accounts of, 88–93, **90**, **92**
      Rutherford's attempt to replicate, 83, 87
   and the Nobel prize in chemistry, 28, 78
   producing neon from radium emanations, 2–3
   Soddy working with, 37, 46, 65
      with radium, 51–56, 59, 66, 68, 71, 77, 84, 93, 112–113, 130, 332, 333
   use of beta particles to induce transmutations, 113
      use of beta particles to induce transmutations, 117, 118
   working with Egerton, 137
rare (noble) gases, xiii, xv, 28, 40, 95, 96, 119–138, 143, 147, 299, 316, 330, 331. *See also* argon; helium; krypton; neon; radon; xenon
Raunitzky, Michael, 24
Redgrove, Herbert Stanley, 146
   Alchemy: Ancient and Modern (Redgrove), 88
Reeves, Richard, 46
repeatability as proof of science. *See* scientific method and repeatability
Reuters, 306
Richardson, O. W., 161–162
Richardson, William H., 168
Riding, R. W., 145, 326
Riesenfeld, Ernst Hermann, 258, 260–261, 266
Rikagaku Kenkyūjo (RIKEN), 238, 249
Ripka, Georges, 296
Robinson, Percy Lucock, 200, 206–208
Roentgen, Wilhelm, 21–23, **22**, 27, 35, 71, 95, 128, 330
Roosevelt, Franklin D., 336
Royal Academy of Sciences, 59
Royal Astronomical Society, 299
Royal Institution of Great Britain, 151, 292
Royal Society, 43, 151
Royds, Thomas, 2, 83, 87, 332, 333
ruby glass, used in transmutation experiments, 241, **242**, 243–244
Rudolph II (emperor), 9–10
Runge, Carl David Tolmé, 237–238
Rutherford, Ernest, **36**, 249, 272, 330, 331, 332, 333
   "Artificial Disintegration of the Elements" (Rutherford), 196
   attempt to replicate Ramsay's man-made transmutations, 83, 87
   being given (or taking) unwarranted credit, xvi, 57–70, 93, 175, 273–296
   Blackett's nitrogen-to-oxygen transmutation, Rutherford's critique, 291–293
   "Collision of Alpha Particles With Light Atoms. IV. An Anomalous Effect in Nitrogen" (Rutherford), 277, 283
   discovery and identification of tritium, 127
   disintegration work
      claiming nitrogen-to-oxygen example of, 273–274, 279, 280–282,

283–284, 287, 289, 291, 297
and concept of natural disintegration, 35–46, 57, 68, 69, 77, 89, 331
disintegration vs. transmutation in thorium-to-argon experiment, 43–44
first man-made disintegration of a stable atom, 147–155, 176, 213, 270, 273–274
follow up to 1919 experiments, 196
high-energy disintegration experiment, 297
man-made disintegration work in 1919. *See* alpha (particle, emission); Rutherford, Ernest, proton emission after bombardment
man-made disintegration work in 1925, 281, 282, 289, 294. *See also* Blackett, Patrick Maynard Stewart, nitrogen-to-oxygen transmutation
natural disintegration of unstable elements, 290
observation of natural nuclear disintegration, 35, 37–38
publications on nitrogen disintegration, 327
on emanations, 2–3, 65, 66–67
"The Energy in the Atom. Can Man Utilize It?" (Rutherford), 109
failing to see a future for nuclear energy, 108–110, **109**
fear that work would be seen as alchemy, 37, 42, 57–58, 60, 68, 148

"General Theoretical Considerations" (Rutherford and Soddy), 58
Law of Radioactive Change (Rutherford and Soddy), 56, 58–59, 60, 61–62, **62**, 64–65, 66, 69, 332
model of the atom, 238, 244, 334
myths
about transmuting metals, **177**
of accomplishing first transmutation of elements, 317
of transmuting hydrogen into helium, 154–155, 175
of transmuting lead into gold, 157, 159, 162–165
of transmuting nitrogen into hydrogen, 157–162
of transmuting nitrogen into oxygen, xvi, 147, **148**, 150–153, 160, 176–177, **177**, 273–287, **275**, 289, 292–294, 295, 319, 334. *See also* Blackett, Patrick Maynard Stewart
The Newer Alchemy (Rutherford), 293–294
and the Nobel prize in chemistry, 44, 46, 57, 59, 65, 195, 290
proton emission after bombardment
work done in 1919, 89, 147, 150–154, 158–159, 176–177, 213, 217, 273–274, 276–279, 283–285, 307–308, 317, 334
work done in 1925, 274–275, **275**, 281, 282, 289, 290, 293–294
radiation, study of, 28
Law of Radioactive Change (Rutherford and

Soddy), 56, 58–59, 60, 61–62, **62**, 64–65, 66, 69, 332
"Radioactive Change" (Rutherford and Soddy), 58, 65
Radio-Activity (Rutherford), 68–69, 332
radium, study of, 2–3, 34
    Rutherford and Soddy's work with, 46, 50, 57–58, 61, 64, 66–67, 89, 93, 150, 290
    seeing protons as hydrogen particles, 155, 158, 177, 284, 294, 317, 334
    thorium experiments with Soddy, 35, 37, 38, 39–45, **41**, 50, 51, 58, 61–63, **62**, 64, 66, 68, 331, 332
    use of individual alpha rays, 185
    Wendt and Irion's experiments, Rutherford's critique, **192**, 192–195, 197–198, 209, 210, 244, 249, 317

San Mateo (California) Times (newspaper), 255
Saturn Model of the atom (Nagaoka), **241**, 244
scattering anomaly, 333
Schleede, Arthur, 258, 260
Schultheis, D., 323
Science (journal), 9, 26–27, 72, 192–193, 196, 197, 203, 307, 313
Science Abstracts: Physics (publication), 208–209
Science Illustrated (publication), 181
Science Service (journal), 231–233, 245–246, 256, 257–258, 263
Scientific American (journal), 222, 243, 259–260, 261–262
    on gold-to-mercury transmutation, 262–263, 269
    "The Retreat of the Modern Alchemists" (an obituary for alchemy), 270, 271–272
    sponsoring efforts to recreate Miethe-Stammreich transmutation, 222, 226–234, **227**, **230**, 237, 258–260, 261–262, 271, 297, 318
scientific method and repeatability, 81–82, 112, 137–138
    efforts to recreate Chicago exploding-wire experiment of Wendt and Irion, 198, 199–211, 295, 317
    efforts to recreate hydrogen-to-helium and neon transmutation (Collie and Paterson), 122–123, 136, 333
    efforts to recreate hydrogen-to-helium transmutation (Paneth and Peters), 313
    efforts to recreate Miethe-Stammreich transmutation of gold, 222–225, **223**, **224**, 226–235, **227**, **230**, 231–233, **233**, 237, 250, 251–279, **259**, 318
    efforts to recreate Ramsay's man-made transmutations, 2, 83, 85–86, 87, 91, 129–130, 277
The Scientific Monthly (journal), 88, 216
scintillation method, 280, 281, 283–284
Sclove, Richard E., 38, 39–40
Scranton Republican (newspaper), **109**
Shadduck, Hugh Allen, 295–296
Sheldon, Harold Horton, 229–230, **230**, 233, **233**, 233, **233**, 258–262, 263, 265, 269
Shimizu, Takeo, 280
Siemens and Halske Works [Aktiengesellschaft], 251–252, 253, 258, 265, 266, 318
Sinclair, Upton
    The Goose Step — a Study of American Education (Sinclair), 132
Skinner, Clarence, 129, 332

Slosson, Edwin E., 231–232
Smith, Oscar Franklin
  Matter and Energy: An Introduction by Way of Chemistry and Physics to the Material Basis of Modern Civilization, Vol. 1 (Wendt and Smith), 181
Smith, Sinclair, 200, **201**, 202, 209
Smithells, Arthur, **141**, 141–142
Smits, Arthur, 256–258, **257**, 263, 265, 271, 272, 306, 318, 335
Society of Chemical Industry, 78
Soddy, Frederick, 34, **36**, **102**, 298, 325, 330, 331, 332, 333
  accepting Collie and Patterson claims of transmutation, 145
  "Alchemy and Chemistry" (Soddy), 39, 41, 42
  and the concept of isotopes, 103
  conductivity of certain gases (Soddy and MacKenzie), 95–96, 112–113
  Disintegration Theory, 42, 44, 57, 59–60, 65–69, 77, 332
  foreseeing potential of nuclear power, 95–104
  "General Theoretical Considerations" (Rutherford and Soddy), 58
  Interpretation of Radium (Soddy), 97–100
  and natural nuclear disintegration, 35, 35–46, 37–38, 290, 331
  and the Nobel prize in chemistry, 69, 103
  radiation, study of, 28
    Law of Radioactive Change (Rutherford and Soddy), 56, 58–59, 60, 61–62, **62**, 64–65, 66, 69, 332
  "Radioactive Change" (Rutherford and Soddy), 58, 65
  Radio-Activity: An Elementary Treatise From the Standpoint of the Disintegration Theory (Soddy), 69, 332
  Soddy plaque at Glasgow University, **70**
  thorium experiments with Rutherford, 35, 37, 38, 39–45, 41, 50, 51, 58, 61–63, **62**, 64, 66, 68, 331, 332
    working with Ramsay, 37, 46, 65
    using radium, 51–56, 59, 66, 68, 71, 77, 84, 93, 112–113, 130, 332, 333
sodium, 2, 78, 83, 85, 332
Souther, Randy, 329
spectral lines, 116, 188, 239, 240, 301, 318
spectrometer/spectroscope, 52–53, **54**, 133, 194, 269, 290, 301
  full spectrum of visible light, **53**
  mass spectrometer, 125
  optical spectrometer, **52**
spectroscopy, 112, 133, 167, 290
  optical emission spectroscopy, 112, 316
  Raman spectroscopy, 218
spontaneous nuclear transmutations, 77
SRI International laboratory, 96
Stammreich, Hans, **219**
  mercury-to-gold transmutations, 217–229, 237, 252–256, 257, 297, 306, 317–318, 334
    efforts to recreate or discredit research, 222–225, **223**, **224**, 226–235, **227**, **230**, 231–233, **233**, 237, 250, 251–279, **259**, 297, 318
    Siemens and Halske acquiring rights to research, 251–252
Standard Model physics, 337–338
stars. *See* sun and stars
Stephenson, George Edward, 200, 206–208

378 • INDEX

Sternglass, Ernest, 336
Stewart, Alfred Walter, 123–125, 145, 148–149, 150–151, 279
   Chemistry and Its Borderland (Stewart), 123
   Recent Advances in Physical and Inorganic Chemistry (Stewart), 88, 123, 149
Strassmann, Fritz, 98, 336
strong force, 4, 33, 337. *See also* electromagnetic force
strong-interaction collisions, 334
Strutt, John William, 28, 136
Strutt, Robert John, 3, 28, 136, 325
subatomic particles, 29–34. *See also* electrons; neutrons; protons
sun and stars, 28, 160
   process that powers, 335, 336
   Wendt on, 185–187, 189–190
Sunday Times (London), 146
Szilárd, Leó, 335, 336

Taylor, Harold John, 336
Tech Engineering News (publication), 262
thallium, xv, 263, 268, 318
theory of special relativity (Einstein), 50, 78, 98, 109, 203, 213, 332
thermal neutron, 336
Thomasville (Georgia) Times-Enterprise (newspaper), 180–181
Thomson, Joseph John (J. J.), 2–3, 113, **126**, 148, 325, 331, 332, 333
   attempt to develop a model of the atom, 238
   challenging Collie-Patterson experiments, 125–129, 145, 175–176, 289–290, 316
   discovery of the electron, 26, 125, 130
   Hirshberg's use of Thomson's work to attack Ramsay, 130–131
   neon experiments, 113, 126–127, 129, 132, 137–138, 143
   and the Nobel prize in physics, 125, 289, 290
   occlusion hypothesis of, 113, 126, 127–129, 133, 143, 175–176, 289
   disproving of, 113, 143, 176, 194, 204
   work with neon, 175–176
   and X_3 (tritium), 75, 113, 126–129, 143
thorium, 2
   nucleosynthetic reaction network of change for, **63**
   Ramsay's work with, 3
   Rutherford and Soddy's work with, 35, 37, 38, 40–45, 61–63, **62**, 64, 66, 68, 331, 332
   thorium-to-argon transmutation, 39–42, **41**, 50, 51, 58
   thorium-to-helium transmutation, 38
   thorium-to-thorium-X transmutation, 37, 38, 43–44, 58–59, **62**, 62–63, 64, 331, 332. *See also* radon
Tiede, Ehrich, 258, 260
Tiffany and Company, 249
Tilden, William A., 78
   Sir William Ramsay (Tilden), 88
Times Literary Supplement (newspaper), 50, 66
Times of London (newspaper), 108, **109**, 159, 160, 193
Tokyo Imperial University, 238, 248, 263
Tokyo Institute of Physical and Chemical Research, 201, 245
transformation. *See* transmutation
transformer oil, 240, 243, 318, 335
transmutation, 332. *See also* man-made transmutation; media coverage of transmutation experiments; natural nuclear transmutation; nuclear transmutation
   and alchemy, 1–4, 5–20
   changes in protons causing, 32–33

credit for first confirmed transmutation, 273–296
versus disintegration, 38, 43–44. See also disintegration
early research rediscovered, 71–76
end of early transmutation era (1925-1927), 251–272
energy involved in, 3, 33, 47, 50, 78, 84, 98–100, 101, 103–104
and exploding wires, 167–211
of gold, 157–165, 213–250
of helium, 297–314
interest in declining (1913-1927), 139–146
mass deficit when transmuting upward on the periodic table, 216–217
and rare gases, 111–138
spontaneous nuclear transmutations, 77
transmutations caused by high-energy alpha particles, 272, 290
Wendt calling it decomposition, 183, 190–191
without a radioactive source, xiii
Travers, Morris W., 88, 323
  A Life of Sir William Ramsay (Travers), 88
Trenn, Thaddeus, 88, 93, 149–150, 290
tritium, xv, 75, 113, 127, 143, 176, 316, 333, 335, 338
tungsten
  use of to make gold, 240–241, 243, 318, 335
  Wendt and Irion experiments with, 175–198, **182**, 317, 334
    efforts to repeat Wendt-Irion experiments, 192–198, 199–211, 249, 295, 317

ULMN. *See* ultra-low-momentum neutron (ULMN)

ultra-low-momentum neutron (ULMN), 338
United Kingdom (U.K.), energy sources in, **106, 108**
  Ramsay foreseeing depletion of coal reserves in, 105–108
United Nations Educational, Scientific and Cultural Organization (UNESCO), 181
United Press, 255
University College, London, 68, 82, **83**, 93, 118–119, 123, 137, 142
University of Amsterdam, 256–257, 258, 318
University of Berlin, 258, 260, 263, 266, 297, 311
University of Cambridge, 113, 125, 130, 175, 280. *See also* Cavendish laboratory
University of Chicago, 131, 202, 210, 229–230, 232, 234, 235, 255, 258, 295, 318
  exploding-wire research at, 167–198, 199–200, 244
University of Vienna, 298
unstable isotopes and elements, 55, 127, 235, 244, 290, 331, 332, 338, 341
uranium, 2, 32, 47, 106, 336
  disintegration of, 103, 134
  and radiation, 28, 48, 49, 61, 99, 331
    uranium salts, 23, 24, 25, 28, 330
  uranium oxide, 99
  uranium-to-uranium-X, 44, 58–59
Urutskoev, Leonid, 174, 208
U.S. Department of Energy, **148**
U.S. Geological Survey on artificial gold, 163–164
Usher, F. L., 3
U.S. National Bureau of Standards, 215
  Bibliography of Scientific Literature Relating to Helium, 326

U.S. National Research Council, 209
Vienna Institute for Radium Research, 298
Villard, Paul, 28, 331

Walton, Ernest, 335
Watson, Eleanor M., 167, 201
weak force and weak interactions, 4
    electroweak interactions, xvii, 4, 337
Wendt, Gerald L., xv, 80, 176, 178–181, **179**, 244, 334
    The Atomic Age Opens (Wendt), 181
    Atomic Energy and the Hydrogen Bomb (Wendt), 181, 191
    Chemistry: The Sciences, a Survey Course for Colleges (Wendt, ed.), 181
    Chicago exploding-wire experiment, 175–198, **205**, 317, 334
        criticisms and efforts to repeat, 192–198, 199–211, 249, 295, 317
    "Decomposing the Atom" (Wendt), 183
    Matter and Energy: An Introduction by Way of Chemistry and Physics to the Material Basis of Modern Civilization, Vol. 1 (Wendt and Smith), 181
    Nuclear Energy and Its Uses in Peace (Wendt), 181
    The Prospects of Nuclear Power and Technology (Wendt), 181
    Science for the World of Tomorrow (Wendt), 181, 210–211
    You and the Atom (Wendt), 179, 181
Whyte, Lorna, 329
Wick, Gian-Carlo, 334, 336
Widom, Allan, 4, 244, 320
Widom-Larsen theory, 320, 338
Wilson, Charles Thomson Rees, 285
Wilson cloud method, 274, 280, 281, 282, 285–287, **286**, 290, 296
Winchester, George, **131**, 131–133, 137–138, 201, 316, 325, 333
wire, exploding. *See* exploding-wire phenomenon
World Scientific, 283
Wright, Joseph, portion of a painting of, **xviii**

X_3, 127, 128, 129, 143. *See also* tritium
xenon, 28, 331
Xia, Ji-xing, 323
X-rays, **22**, 25, 119, 330, 332
    discovery of, 21–23, 35, 95, 128
    X-ray photos, 171, **172**
    X-ray tubes, 121, 194–195, 333

Yajna-sala. *See* Birla Temple (New Delhi)
Yang, Jian-yu, 323
Yasuda (gold assaying expert), 241
Youth's Companion (magazine), 158–159
YouTube video showing exploding-wire experiment, 172

Zeitsch, J., 326

About the Author

Steven B. Krivit lives in San Rafael, California, and is an investigative science journalist and international speaker. He studied industrial design at the University of Bridgeport (Connecticut) and completed his bachelor's degree in business administration and information technology at National University (Los Angeles). He was a computer network systems engineer until 2000, when he became curious about low-energy nuclear reaction (LENR) research. He founded the *New Energy Times* Web site and online news service to share what he learned. By 2016, he had spoken with nearly all the scientists who were involved in the field. He has lectured nationally and international to scientific as well as lay audiences. He has advised the U.S. intelligence community, the U.S. Library of Congress, members of the Indian Atomic Energy Commission and the interim executive director of the American Nuclear Society. He is the leading author of review articles and chapters about LENRs, including invited papers for the Royal Society of Chemistry (2009), Elsevier (2009 and 2013) and John Wiley & Sons (2011). He was an editor for the American Chemical Society 2008 and 2009 technical reference books on LENRs and editor-in-chief for the 2011 Wiley *Nuclear Energy Encyclopedia.*

Krivit was the first science journalist to publicly identify and teach the distinctions between the unproven theory of "cold fusion" and the experimentally confirmed neutron-catalyzed LENRs. He did so in 2008 at the 236th national meeting of the American Chemical Society. His chapters in the *Elsevier Encyclopedia of Electrochemical Power Sources* were the first chapters on LENRs in a print encyclopedia.

Other Volumes in This Series

Hacking the Atom: Explorations in Nuclear Research, Vol. 1

This book shows, for the first time, why low-energy nuclear reaction phenomena are not the result of fusion, why they are the result of nuclear processes, and why they can now be explained by a feasible theory. The theory does not conflict with existing physics but expands scientific knowledge and reveals a new field of nuclear science.

Fusion Fiasco: Explorations in Nuclear Research, Vol. 2

This book tells the behind-the-scenes story of the 1989-1990 fusion fiasco, one of the most divisive scientific controversies in recent history. It explains how credible experimental low-energy nuclear reactions research emerged from the erroneous idea of "cold fusion."

For More Information
www.stevenbkrivit.com

CPSIA information can be obtained
at www.ICGtesting.com
Printed in the USA
FFOW02n0151311217
44212021-43660FF